BRITISH PYRALID MOTHS

Frontispiece 1. *Agriphila tristella* (D.&S.), 2. *Catoptria pinella* (Linn.) (CRAMBINAE); 3. *Donacaula forficella* (Thunb.) (SCHOENOBIINAE); 4. *Scoparia ambigualis* (Treit.) (SCOPARIINAE); 5. *Elophila nymphaeata* (Linn.) (NYMPHULINAE); 6. *Evergestis pallidata* (Hufn.) (EVERGESTINAE); 7. *Cynaeda dentalis* (D.&S.) (ODONTIINAE); 8. *Anania funebris* (Ström), 9. *Nomophila noctuella* (D.&S.) (PYRAUSTINAE); 10. *Pyralis farinalis* (Linn.) (PYRALINAE); 11. *Galleria mellonella* (Linn.) (GALLERIINAE); 12. *Dioryctria abietella* (D.&S.) (PHYCITINAE). Figs 1–3, 5–7, 9–12, × 2; Fig. 4, × 2·5; Fig. 8, × 1·5. *Photos:* M.W.F. Tweedie

BRITISH PYRALID MOTHS

A Guide to their Identification

Barry Goater

Illustrated by Geoffrey Senior and
Robert Dyke

HARLEY
BOOKS

Harley Books (B. H. & A Harley Ltd.),
Martins, Great Horkesley,
Colchester, Essex, CO6 4AH, England

Text set in Linotron 202 Plantin by
Rowland Phototypesetting Ltd.,
and printed by St Edmundsbury Press,
Bury St Edmunds, Suffolk

Colour originated by Adroit Photo Litho Ltd., Birmingham,
and printed by Jolly & Barber Ltd., Rugby

Bound by Hunter & Foulis Ltd., Edinburgh

British Pyralid Moths

ISBN 0 946589 08 9

Contents

Foreword

It is almost exactly 100 years since the appearance of Leech's *British Pyralides*, and nearly 35 since that of Beirne's *British Pyralid and Plume Moths*. The time is ripe for publication of a new manual incorporating recent discoveries and utilizing modern resources of colour photography and reproduction. Barry Goater, a proven author and an assiduous student of the British fauna, has shown himself well able to supply the need. His clear descriptions and concise notes on variation, life histories, habits, seasonal occurrence and geographical distribution are complemented by an excellent set of colour plates, prepared from photographs, and line drawings needed to illustrate structural or pattern differences between some species.

The rapid progress of our knowledge of pyrales is shown by the fact that, whereas Beirne knew 174 British species, the present work lists 208. This nearly twenty per cent increase comprises previously unrecognized species, new immigrants, migrants, accidental introductions, and an interesting set of exotics established on aquatic plants in greenhouses. There is progress, too, in the classification: modern changes in subfamily and generic concepts and nomenclature are reflected in Mr Goater's text. Species names, also, have been changed to accord with present views on distinguishing characters and with recent studies of nomenclature and of the identity of the older type-specimens.

Though Mr Goater has provided us with an excellent compendium of present knowledge of British Pyralidae, his book will perhaps be of greatest value to seekers of yet undiscovered information. Not only does he give us a base line from which journeys of exploration can start, but where possible he also provides us with signposts telling us where we ought to look and what knowledge we most urgently lack. Life histories, host plants, habits, seasonal distribution of adults and early stages, resident, migrant or other status, and geographical variation in appearance and biology are among the aspects that offer rich opportunities in pyralid biology. I would like to put in a personal plea for comparative structural studies of eggs, larvae and pupae, largely neglected in the British fauna, though begun in some others.

Whatever these future investigations may reveal, Mr Goater's work will stand as a landmark in publications on the British fauna, and we should all feel indebted to him for his contribution.

Eugene Munroe, PHD, FRSC, FESC, FRES

August, 1985

Dunrobin, Ontario, Canada

To my Colleagues and Pupils
past and present at
Haberdashers' Aske's School, Elstree

Preface

In accepting responsibility for writing the text of this guide to the identification of British Pyralidae, I am acutely aware that what B. P. Beirne wrote in the first paragraph of the preface to his book on the British pyralid and plume moths is as true today as it was in 1952. These moths remain comparatively neglected: they are less well-known to amateur lepidopterists than most other families of Lepidoptera. Indeed I believe it to be true that no British field naturalist ranks the Pyralidae first among the groups he studies. The late H. C. Huggins came closest in recent years to being an authority, and it is unfortunate that he did not write this book. My purpose is to stimulate interest by providing an up-to-date aid to the identification of the British pyralid moths and to expose some of the gaps in our knowledge in the hope that more will be known by the time the appropriate volume of *The Moths and Butterflies of Great Britain and Ireland* comes to be written. My burden has been lightened beyond measure by the expertise of Mr G. B. Senior, who has photographed in colour all the species on the British list, and by the artistic skills of Mr R. Dyke, who has made drawings from preparations carefully made by the Rev. D. J. L. Agassiz, and also of Mr T. H. Freed who drew the generalized pyralid and genitalia figures for the introduction to the family. Generous assistance has been rendered by others mentioned below, but I alone am responsible for the shortcomings of the text. The decision to omit the Pterophoridae or plume moths from this work was also mine – I hope this may prompt others to take up the pen on their behalf!

In preparing the text, I have received encouragement and factual assistance from fellow entomologists too numerous to name. The draft was read in full by Messrs Agassiz, J. M. Chalmers-Hunt and E. C. Pelham-Clinton, and I am most grateful to these gentlemen for their helpful criticism and comment. Dr J. D. Bradley and Dr E. G. Munroe have advised comprehensively concerning nomenclature, and especially on changes proposed since Kloet and Hincks' check list. The loan of specimens for photography by the British Entomological and Natural History Society and by Mr Agassiz, Dr M. W. Harper, Mr B. Skinner and Col. D. H. Sterling is gratefully acknowledged. The Trustees of the British Museum (Natural History) have been most kind in allowing free and extensive study of the collections and permitting me to borrow many specimens, including some extreme rarities,

9

to be photographed. Within the Department of Entomology, I have received friendly co-operation and encouragement from Dr K. Sattler and Mr P. E. S. Whalley, and especially from Mr M. Shaffer. Mr B. H. Harley has been tireless in checking the references and advising in many other ways, helped by Mr A. R. Waterston. Mr M. W. F. Tweedie kindly allowed me to use his excellent colour photographs of pyralids in the wild for the frontis-piece and the jacket. My final pleasure is to thank my wife, Jane, for her constant encouragement and active help.

Barry Goater
Bushey, Hertfordshire

December, 1985

Introduction

The object of this book is to help naturalists and conservationists to identify the species of British Pyralidae. The majority can be recognised by comparing them with the coloured illustrations which contain all of the 208 species currently on the British list. The text aims to give a concise description of each species and its more striking races or forms, and to emphasise differences between similar species. Line-drawings are included in a few instances where it is thought that these will be helpful. In a small number of cases, reference to the structure of the genitalia (text figs 2a–c, p. 21) is essential for identification, and drawings of the male and female structures are given for all the difficult species, but for other species see Pierce & Metcalfe (1938). It was not thought necessary to provide illustrations of the genitalia of all the fauna, nor was it possible to produce a satisfactory key for the identification of species without introducing technicalities inaccessible to the average amateur lepidopterist. Nomenclature follows closely that of *A Recorder's Log Book or Label List of British Butterflies and Moths* (Bradley & Fletcher, 1979; 1983), but certain modifications have been made in accordance with recent work. Habits, life history and British distribution are given in outline only: for fuller details of these, the reader is recommended to consult the works of Barrett (1904), Meyrick (1928) and Beirne (1952), or to await Volume 6 of *The Moths and Butterflies of Great Britain and Ireland* (eds Heath & Emmet, 1976–).

Of the 208 British species, about 140 can be regarded as native, established outdoors in the wild. These will be of greatest interest to the conservationist, for all have their particular habitat preferences and some of these are subject to pressures which could lead to a reduction in their numbers or even loss from our fauna. A number of the Crambinae, for instance, are characteristic of particular kinds of grassland. Our diminishing calcareous downlands support several interesting pyraustines, and other species occur only on heaths and moors, in wetlands or in particular kinds of woodland. A few species are, or appear to have become, exceedingly local and rare. *Eurhodope cirrigerella* (Zincken) may be extinct; *Pempelia obductella* (Zeller) and *Agrotera nemoralis* (Scopoli) are extremely local and scarce in south-east England and may merit special conservation measures, while the presence of *Salebriopsis albicilla* (Herrich-Schäffer), found only in woods containing small-leaved lime in the Wye Valley, is just one reason for ensuring that such woodland is never destroyed. The needs of *Gymnancyla canella* (Denis & Schiffermüller) and *Melissoblaptes zelleri* (de Joannis) should be taken into account by those providing for increased tourism on the coast. On the other hand, the evident establishment and spread of such as *Phlyctaenia perlucidalis* (Hübner) and *Dioryctria schuetzeella* Fuchs need sensitive monitoring.

About 25 pyralid species are migrants: some are regular and common most years, and among the best known are *Nomophila noctuella* (Denis & Schiffermüller) and *Udea ferrugalis* (Hübner). Some establish themselves as breeding species for a few years: examples are *Udea fulvalis* (Hübner), *Margaritia sticticalis* (Linnaeus) and *Ancylosis oblitella* (Zeller). The rest are uncommon to exceedingly rare, and some have been recorded only once. The number of species in this category is added to from time to time. Their doings are beyond the influence of conservationists but are of great interest to the fortunate field naturalist who encounters them.

About three dozen pyralid species are pests of stored products (see Baker, 1976) or are introduced from time to time with imported plant material and can survive only in special conditions, or live predominantly indoors in farm buildings. Pests such as *Ephestia kuehniella* Zeller are always with us, others appear to be chance imports, and several may be overlooked because no entomologist ever goes their way. One curious group of adventives is the Nymphulinae which appear in greenhouses specializing in tropical water-plants for aquarists. Two species in this broad category which might merit conservation measure on their behalf are the barn-inhabiting Pyralinae, *Pyralis lienigialis* (Zeller) and *Aglossa caprealis* (Hübner).

The remaining species are of doubtful status but include those which might conceivably be resident in Britain such as *Catoptria verellus* (Zincken) and *Metaxmeste phrygialis* (Hübner). It is up to the field lepidopterist to find them.

Check List of British Pyralidae

CRAMBINAE
Euchromius ocellea (Haworth, 1811)
Chilo phragmitella (Hübner, 1805)
Acigona cicatricella (Hübner, 1824)
Calamotropha paludella (Hübner, 1824)
Chrysoteuchia culmella (Linnaeus, 1758)
  =*hortuella* (Hübner, 1796)
Crambus pascuella (Linnaeus, 1758)
C. leucoschalis Hampson, 1898
C. silvella (Hübner, 1813)
C. uliginosellus Zeller, 1850
C. ericella (Hübner, 1813)
C. hamella (Thunberg, 1788)
C. pratella (Linnaeus, 1758)
  =*dumetella* (Hübner, 1813)
C. lathoniellus Zincken, 1817
  =*nemorella* auctt.
C. perlella (Scopoli, 1763)
Agriphila selasella (Hübner, 1813)
A. straminella (Denis & Schiffermüller, 1775)
  =*culmella* auctt.
A. tristella (Denis & Schiffermüller, 1775)
A. inquinatella (Denis & Schiffermüller, 1775)
A. latistria (Haworth, 1811)
A. poliellus (Treitschke, 1832)
A. geniculea (Haworth, 1811)
Catoptria permutatella (Herrich-Schäffer, 1848)
  =*myella* auctt. *nec* Hübner, 1796
C. osthelderi (de Lattin, 1950)
C. speculalis Hübner, 1825
C. pinella (Linnaeus, 1758)
C. margaritella (Denis & Schiffermüller, 1775)
C. furcatellus (Zetterstedt, 1840)
C. falsella (Denis & Schiffermüller, 1775)
C. verellus (Zincken, 1817)
C. lythargyrella (Hübner, 1796)
Chrysocrambus linetella (Fabricius, 1781)
  = *cassentiniellus* (Herrich-Schäffer, 1848)
C. craterella (Scopoli, 1763)
  =*rorella* (Linnaeus, 1767)
Thisanotia chrysonuchella (Scopoli, 1763)
Pediasia fascelinella (Hübner, 1813)
P. contaminella (Hübner, 1796)

P. aridella (Thunberg, 1788)
=*salinellus* (Tutt, 1887)
Platytes alpinella (Hübner, 1813)
P. cerussella (Denis & Schiffermüller, 1775)
Ancylolomia tentaculella (Hübner, 1796)

SCHOENOBIINAE
Schoenobius gigantella (Denis & Schiffermüller, 1775)
Donacaula forficella (Thunberg, 1794)
D. mucronellus (Denis & Schiffermüller, 1775)

SCOPARIINAE
Scoparia subfusca Haworth, 1811
=*cembrella* auctt.
S. pyralella (Denis & Schiffermüller, 1775)
=*arundinata* (Thunberg, 1792)
=*dubitalis* (Hübner, 1796)
S. ambigualis (Treitschke, 1829)
S. basistrigalis Knaggs, 1866
S. ancipitella (de la Harpe, 1855)
=*ulmella* Knaggs, 1867
Dipleurina lacustrata (Panzer, 1804)
=*crataegella* auctt.
=*centurionalis* sensu Beirne, 1952
Eudonia pallida (Curtis, 1827)
E. alpina (Curtis, 1850)
=*borealis* (Tengström, 1848) *nec* (Lefebvre, 1836)
E. murana (Curtis, 1827)
E. truncicolella (Stainton, 1849)
E. lineola (Curtis, 1827)
E. angustea (Curtis, 1827)
E. delunella (Stainton, 1849)
=*vandaliella* (Herrich-Schäffer, 1851)
=*resinella* auctt.
E. mercurella (Linnaeus, 1758)

NYMPHULINAE
Elophila nymphaeata (Linnaeus, 1758)   Brown China-mark
E. difflualis (Snellen, 1882)
=*enixalis* (Swinhoe, 1885)
E. melagynalis (Agassiz, 1978)
E. manilensis (Hampson, 1917)
Synclita obliteralis (Walker, 1859)
Nymphula stagnata (Donovan, 1806)   Beautiful China-mark
Parapoynx stratiotata (Linnaeus, 1758)   Ringed China-mark
P. obscuralis (Grote, 1881)
P. diminutalis (Snellen, 1880)
P. fluctuosalis (Zeller, 1852)
P. crisonalis (Walker, 1859)
=*stagnalis* sensu Agassiz, 1981
Oligostigma polydectalis Walker, 1859
O. angulipennis Hampson, 1891

O. bilinealis Snellen, 1876

Cataclysta lemnata (Linnaeus, 1758)   Small China-mark

ACENTROPINAE

Acentria ephemerella (Denis & Schiffermüller, 1775)   Water Veneer
  =*nivea* (Olivier, 1791)

EVERGESTINAE

Evergestis forficalis (Linnaeus, 1758)   Garden Pebble

E. extimalis (Scopoli, 1763)

E. pallidata (Hufnagel, 1767)

ODONTIINAE

Cynaeda dentalis (Denis & Schiffermüller, 1775)

Metaxmeste phrygialis (Hübner, 1796)

GLAPHYRIINAE

Hellula undalis (Fabricius, 1781)   Old World Webworm

PYRAUSTINAE

Pyrausta aurata (Scopoli, 1763)

P. purpuralis (Linnaeus, 1758)

P. ostrinalis (Hübner, 1796)

P. sanguinalis (Linnaeus, 1767)

P. cespitalis (Denis & Schiffermüller, 1775)

P. nigrata (Scopoli, 1763)

P. cingulata (Linnaeus, 1758)

Margaritia sticticalis (Linnaeus, 1761)

Uresiphita polygonalis (Denis & Schiffermüller, 1775)
  =*limbalis* (Denis & Schiffermüller, 1775)
  =*gilvata* (Fabricius, 1794)

Sitochroa palealis (Denis & Schiffermüller, 1775)

S. verticalis (Linnaeus, 1758)

Paracorsia repandalis (Denis & Schiffermüller, 1775)

Microstega pandalis (Hübner, 1825)

M. hyalinalis (Hübner, 1796)

Ostrinia nubilalis (Hübner, 1796)   European Corn-borer

Eurrhypara hortulata (Linnaeus, 1758)   Small Magpie

Perinephela lancealis (Denis & Schiffermüller, 1775)

Phlyctaenia coronata (Hufnagel, 1767)
  =*sambucalis* (Denis & Schiffermüller, 1775)

P. perlucidalis (Hübner, 1800–09)

P. stachydalis (Germar, 1822)

Mutuuraia terrealis (Treitschke, 1829)

Anania funebris (Ström, 1768)
  =*octomaculata* (Linnaeus, 1771)

A. verbascalis (Denis & Schiffermüller, 1775)

Psammotis pulveralis (Hübner, 1796)

Ebulea crocealis (Hübner, 1796)

Opsibotys fuscalis (Denis & Schiffermüller, 1775)

Nascia cilialis (Hübner, 1796)

Udea lutealis (Hübner, 1800–09)
  =*elutalis* auctt.

U. fulvalis (Hübner, 1800–09)
U. prunalis (Denis & Schiffermüller, 1775)
=*nivealis* (Fabricius, 1781)
U. decrepitalis (Herrich-Schäffer, 1848)
U. olivalis (Denis & Schiffermüller, 1775)
U. uliginosalis (Stephens, 1834)
=*alpinalis* sensu Ford
U. alpinalis (Denis & Schiffermüller, 1775)
U. ferrugalis (Hübner, 1796)
=*martialis* (Guenée, 1854)
Mecyna flavalis (Denis & Schiffermüller, 1775)
subsp. *flaviculalis* Caradja, 1916
M. asinalis (Hübner, 1818–19)
Nomophila noctuella (Denis & Schiffermüller, 1775)   Rush Veneer
N. nearctica Munroe, 1973
Dolicharthria punctalis (Denis & Schiffermüller, 1775)
Antigastra catalaunalis (Duponchel, 1833)
Maruca testulalis (Geyer, 1832)   Mung Moth
Diasemia reticularis (Linnaeus, 1761)
=*litterata* (Scopoli, 1763)
Diasemiopsis ramburialis (Duponchel, 1833)
Hymenia recurvalis (Fabricius, 1775)
Pleuroptya ruralis (Scopoli, 1763)   Mother of Pearl
Herpetogramma centrostrigalis (Stephens, 1834)
H. aegrotalis (Zeller, 1852)
Palpita unionalis (Hübner, 1796)
Diaphania hyalinata (Linnaeus, 1767)   Melonworm
=*lucernalis* (Hübner, 1796)
Agrotera nemoralis (Scopoli, 1763)
Leucinodes vagans (Tutt, 1890)
Sceliodes laisalis (Walker, 1859)

PYRALINAE
Hypsopygia costalis (Fabricius, 1775)   Gold Triangle
Synaphe punctalis (Fabricius, 1775)
=*angustalis* (Denis & Schiffermüller, 1775)
Orthopygia glaucinalis (Linnaeus, 1758)
Pyralis lienigialis (Zeller, 1843)
P. farinalis (Linnaeus, 1758)   Meal Moth
P. manihotalis Guenée, 1854
P. pictalis (Curtis, 1834)   Painted Meal Moth
Aglossa caprealis (Hübner, 1800–09)   Small Tabby
A. pinguinalis (Linnaeus, 1758)   Large Tabby
A. dimidiata (Haworth, 1809)   Tea Tabby
A. ocellalis Lederer, 1863
Endotricha flammealis (Denis & Schiffermüller, 1775)

GALLERIINAE
Galleria mellonella (Linnaeus, 1758)   Wax Moth
Achroia grisella (Fabricius, 1794)   Lesser Wax Moth
Corcyra cephalonica (Stainton, 1866)   Rice Moth

Aphomia sociella (Linnaeus, 1758)   Bee Moth
Melissoblaptes zelleri (de Joannis, 1932)
Paralipsa gularis (Zeller, 1877)   Stored Nut Moth
Arenipses sabella Hampson, 1901

PHYCITINAE
Anerastia lotella (Hübner, 1810–13)
Cryptoblabes bistriga (Haworth, 1811)
C. gnidiella (Millière, 1867)
Salebriopsis albicilla (Herrich-Schäffer, 1849)
Metriostola betulae (Goeze, 1778)
Trachonitis cristella (Hübner, 1796)
Selagia argyrella (Denis & Schiffermüller, 1775)
Microthrix similella (Zincken, 1818)
Pyla fusca (Haworth, 1811)
Phycita roborella (Denis & Schiffermüller, 1775)
    =*spissicella* (Fabricius, 1777)
Oncocera semirubella (Scopoli, 1763)
Pempelia palumbella (Denis & Schiffermüller, 1775)
P. genistella (Duponchel, 1836)
P. obductella (Zeller, 1839)
P. formosa (Haworth, 1811)
Sciota hostilis (Stephens, 1834)
Hypochalcia ahenella (Denis & Schiffermüller, 1775)
Epischnia bankesiella Richardson, 1888
Dioryctria abietella (Denis & Schiffermüller, 1775)
D. mutatella Fuchs, 1903
D. schuetzeella Fuchs, 1899
Pima boisduvaliella (Guenée, 1845)
Nephopterix angustella (Hübner, 1796)
Pempeliella diluta (Haworth, 1811)
    =*dilutella* auctt.
P. ornatella (Denis & Schiffermüller, 1775)
Acrobasis tumidana (Denis & Schiffermüller, 1775)
A. repandana (Fabricius, 1798)
    =*tumidella* (Zincken, 1818)
A. consociella (Hübner, 1810–13)
Numonia suavella (Zincken, 1818)
N. advenella (Zincken, 1818)
N. marmorea (Haworth, 1811)
Apomyelois bistriatella (Hulst, 1887)
subsp. *neophanes* (Durrant, 1915)
Ectomyelois ceratoniae (Zeller, 1839)   Locust Bean Moth
Mussidia nigrivenella Ragonot, 1888
Eurhodope cirrigerella (Zincken, 1818)
Myelois cribrella (Hübner, 1796)   Thistle Ermine
Gymnancyla canella (Denis & Schiffermüller, 1775)
Zophodia grossulariella (Zincken, 1818)
    =*convolutella* auctt.
Assara terebrella (Zincken, 1818)

Euzophera cinerosella (Zeller, 1839)
E. pinguis (Haworth, 1811)
E. osseatella (Treitschke, 1832)
E. bigella (Zeller, 1848)
Nyctegretis lineana (Scopoli, 1763)
    =*achatinella* (Hübner, 1823–24)
Ancylosis oblitella (Zeller, 1848)
Homoeosoma sinuella (Fabricius, 1793)
H. nebulella (Denis & Schiffermüller, 1775)
H. nimbella (Duponchel, 1836)
Phycitodes maritima (Tengström, 1848)
    =*carlinella* (Heinemann, 1865)
    =*cretacella* (Rössler, 1866)
P. binaevella (Hübner, 1810–13)
P. saxicola (Vaughan, 1870)
Plodia interpunctella (Hübner, 1810–13)   Indian Meal Moth
Ephestia elutella (Hübner, 1796)   Cacao Moth
E. parasitella Staudinger, 1859
subsp. *unicolorella* Staudinger, 1881
E. kuehniella Zeller, 1879   Mediterranean Flour Moth
E. cautella (Walker, 1863)   Dried Currant Moth
E. figulilella Gregson, 1871   Raisin Moth
E. calidella Guenée, 1845   Dried Fruit Moth

Description of British Pyralid Moths

Family PYRALIDAE

The Pyralidae comprise an extensive family of moths, mostly over 15mm wingspan which, on account of their size, might be mistaken by the inexperienced for macrolepidoptera, but which differ from them in the presence of vein CuP in the hindwing; vein Sc+R$_1$ approximates to or anastomoses with vein Rs beyond the cell, then diverges. The presence of tympanal organs at the base of the abdomen is a feature shared only with the Geometroidea; their structure demands detailed study and may afford valuable taxonomic characters. The labial palpi are frequently long, porrect or upturned; the antennae are characteristically long, simple, and quickly become brittle when the dead specimen begins to dry. The legs, too, are often long, slender and brittle. The British species are slender-bodied, mostly small to medium-sized insects of 15–30mm wingspan; the forewing varies in shape from broadly triangular to very narrow, but the hindwing is characteristically ample. In many species, the markings of the type depicted in text fig. 1 (p. 20) are clearly defined, but in others they are obscure or absent. In many British species of the subfamily Phycitinae, the forewing has two dark dots, one above the other, in the approximate position of the reniform stigma, and little other pattern. The Pyralidae are chiefly tropical, and the few subfamilies that are represented in Britain give small indication of the diversity of size, build, colour and posture to be found among tropical species.

The larvae of Pyralidae are typically slender, sparsely setose, active animals which live in spun or rolled leaves or silken tubes and galleries; when disturbed, they run forwards and backwards with equal agility. But their habits are varied and their diets diverse: all parts of plants may be attacked including the cones and bark of gymnosperms, and even bryophytes which most animals seem to find inedible; one group is secondarily aquatic, the larvae showing interesting structural modification in the form of tracheal gills; others are specialists in dried organic material including stored produce intended for human consumption, and many warehouse pests are pyralids. These pest species are usually very easy to breed and can be kept in continuous culture, but others are more tricky to rear. Many species have the annoying habit of overwintering as fully grown larvae in the cocoon in which they will eventually pupate, and often succumb to drought or disease at the hands of the inexperienced or unlucky entomologist, and probably also in the wild. No serious attempt seems to have been made to breed many of the species, especially from the egg, and there is much interesting and rewarding research to be done in this field.

Whereas many pyralid species are active fliers or easily disturbed by day, others are strictly nocturnal; most have characteristic resting postures, ways of flying, courtship behaviour and techniques of oviposition. All too little of this field natural history, so well-documented in other groups, is known for the Pyralidae.

The British species are classified under twelve subfamilies, as follows:

Crambinae (p. 22)
Schoenobiinae (p. 44)
Scopariinae (p. 46)
Nymphulinae (p. 55)
Acentropinae (p. 62)
Evergestinae (p. 63)
Odontiinae (p. 64)
Glaphyriinae (p. 65)
Pyraustinae (p. 66)
Pyralinae (p. 94)
Galleriinae (p. 99)
Phycitinae (p. 103)

The main characters of each subfamily are given in the descriptive text.

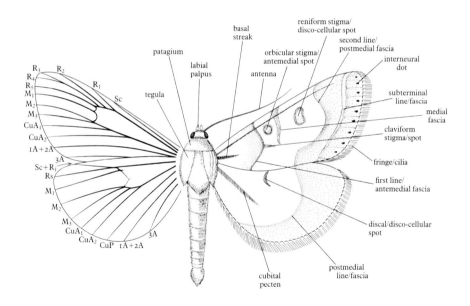

Text figure 1

Generalized pyralid
showing wing venation as seen from below
and wing pattern from above

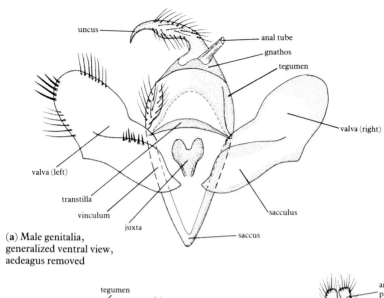

(a) Male genitalia, generalized ventral view, aedeagus removed

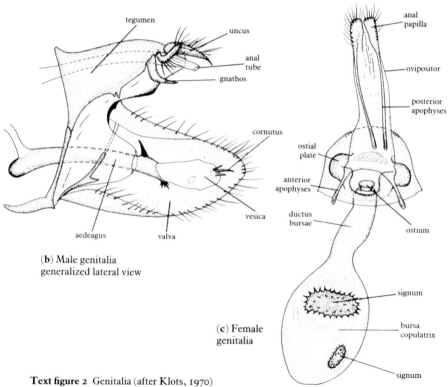

(b) Male genitalia generalized lateral view

(c) Female genitalia

Text figure 2 Genitalia (after Klots, 1970)

Subfamily CRAMBINAE (39 species)

The Crambinae, or grass moths, are familiar to many naturalists. The species most often encountered are those which are disturbed from among long grass on sunny days, fly off with rapid zig-zag flight and settle head downwards on grass-stems a few metres ahead, wings tightly wrapped around the body. They appear larger on the wing when their ample hindwings are displayed. Not all species are so readily put up in daytime, and some can be found only at night. Some are migrants. The larvae feed on roots of grasses and sometimes on moss.

Euchromius ocellea (Haworth)

Pl.1, fig.9

IMAGO Wingspan 16–27mm. Antenna of male ciliate. Forewing of male with semitransparent patch in cell towards base: forewing otherwise pale greyish ochreous, sparsely sprinkled black in terminal quarter, finely strigulate silver, giving the whole wing a characteristic high gloss; two rather broad, slightly sinuate, parallel yellowish cross-lines slightly beyond middle of wing, and a fine, strongly angled pale subterminal line, edged distally between angle and tornus by a narrow white patch which contains several oval, golden metallic terminal spots, each edged basally by two or three intensely black dots; fringe whitish with two narrow blackish dividing lines. Hindwing pearly greyish white with darker veins, narrow brown terminal line and white fringe. This species is distinguished from other British pyralids by the pale coloration and terminal band of black and metallic spots, though there are other *Euchromius* species which resemble it closely, and the genitalia of British taken specimens should be checked.

LARVA Described by Capps (1966) who states that it feeds on corn and milo maize (*Sorghum*) roots and that an association with stored produce and dead vegetation (given by European authorities, including Bleszyński, 1965) is doubtless accidental.

DISTRIBUTION A widespread tropical and subtropical species which is an occasional immigrant to Britain (Skinner, 1982). About thirty records; recorded in every month of the year except April and December, nearly half in January–March and about a quarter in September–October; seven in Hampshire, five in Devon, three in Kent and Cheshire, and scattered records in other counties as far north as Dumfriesshire; two in south Wales and a single record from Co. Cork.

Chilo phragmitella (Hübner)

Pl.1, figs 1(♂), 2(♀)

IMAGO Wingspan male 24–32mm, female 30–40mm. Sexually dimorphic. Labial palpi *c*.4mm long. *Male*. Forewing rather broad, costa arched near base, apex acute, termen full, weakly sinuate; dark or light reddish brown, sometimes suffused fuscous, finely dark streaked along and between veins, often with darker longitudinal stripe from base through small, black discal spot. Hindwing pale brownish white. Cilia of all wings with metallic gloss. *Female.* Forewing longer and narrower, apex drawn to a narrow point and termen very oblique; pale wainscot brown with fine longitudinal streaks and small, point-like discal spot. Hindwing pure white. Males are unmistakable; females could be confused with female Schoenobiinae, but in those species the costa is convex towards apex, whereas in *C. phragmitella* it is distinctly concave. The pecten of long setae which is present along the dorsal margin of the cell of the hindwing in *Chilo* species is not always easy to see even with a hand-lens.

Single-brooded; flies in June and July. The moth rests by day among stems of common reed and other vegetation. After dark, males fly weakly through the reed-beds and come freely to light. Females also come to light, but are more inclined to wander from their breeding-ground.

LARVA In the stem and root-stock of common reed (*Phragmites*) or reed sweet-grass (*Glyceria maxima*), September to June. Pupa in a reed-stem below an exit 'window' gnawed by the larva prior to pupation.

DISTRIBUTION A characteristic species of large reed-beds in England south of Durham, and Wales. Recorded from Adderstonelee Moss, Roxburghshire, 1982; mid-Perthshire in 1959; and Tay Estuary in 1967 (Bland, 1983).

Acigona cicatricella (Hübner)

Pl.1, figs 5(♂), 6(♀)

IMAGO Wingspan male 21–24mm, female 34–38mm. Sexually dimorphic. Labial palpi shorter than head and thorax. *Male.* Forewing broad, squared at apex, cream-coloured, variably irrorate light to dark brown except for costa which usually remains pale; the rounded, dark flecked discal spot usually has a pale, cream-coloured spot or short bar each side of it, with indications of a dark streak beyond these markings as far as subterminal line; a fuscous basal streak dorsal to the cell is broken by pale cream-coloured marks; subterminal line evenly curved, formed of pale dots between the veins each with a darker smudge basally; veins pale in terminal region; a series of blackish terminal interneural dots. Hindwing whitish, weakly suffused light brown in terminal area. *Female.* Forewing longer and narrower, apex more oblique, termen slightly sinuate, pattern similar to that of male but colour usually rich mahogany brown with strongly contrasting broad, pale costa. Hindwing white with narrow light brown terminal line. *Donacaula mucronellus* (Denis & Schiffermüller) (Pl.2, figs 30,31) is smaller, with differently shaped wings, forewing more irrorate, with distinct zone of darker suffusion dorsal to costal stripe.

Flies (in France) in June and July. Comes to light.

LARVA Feeds from May to July in a stem of common club-rush (*Scirpus lacustris*) and pupates in the stem (Bleszyński, 1965).

DISTRIBUTION Female taken at light in Kent in 1951 (Chalmers-Hunt, 1952); and one stated to have been taken flying near Dover, Kent, was exhibited by E. Shepherd at the meeting of the Entomological Society of London in September, 1852 (Stainton, 1855).

Calamotropha paludella (Hübner)

Pl. 1, figs 3(♂), 4(♀)

IMAGO Wingspan 23–29mm. Female larger and plainer than male. Forewing rather broad, whitish or light brownish, weakly suffused fuscous; a blackish basal dot with another nearby, two or three representing the antemedian line, one in disc and an oblique row representing the postmedian line: some or all of these dots frequently absent; fringe chequered brown and pale buff. Hindwing shining white with fine, broken, dark brown terminal line; fringe white. Unmistakable.

Single-brooded; flies in July and August. Takes wing at dusk and flies for about an hour over water and among reeds, then again later at night. Comes to light.

LARVA Feeds in late summer and autumn in a mine in a leaf of reedmace (*Typha latifolia*) or, rarely, lesser reedmace (*T. angustifolia*), burrowing towards base where it hibernates; in spring it mines a dead leaf or stem, sometimes gregariously, until June. Pupa in larval mine.

DISTRIBUTION Very local in large wet fens and marshes and on margins of flooded gravel-pits and broads in Norfolk, Suffolk, Cambridgeshire, Essex, Kent, Hampshire, Isle of Wight and Dorset; occasional wanderers turn up elsewhere. Probably overlooked: adults tend to stay amongst the foodplants, where they can sometimes be found in profusion.

Chrysoteuchia culmella (Linnaeus)

= *hortuella* (Hübner)

Pl. 1, figs 7,8

IMAGO Wingspan 20–24mm. Forewing creamy white more or less suffused with leaden grey; in darker specimens (fig.8), the veins remain conspicuously pale except towards costa; in pale specimens (fig.7), a fine, oblique postmedian line, elbowed near costa, is present; a dark, pale-edged, less oblique but more strongly angled subterminal line is constantly present. Hindwing smoky. Both wings with highly golden metallic cilia. Unmistakable: *Agriphila straminella* (Denis & Schiffermüller) (Pl. 1, figs 23–25) is smaller and lacks cross-lines; *Pediasia contaminella* (Hübner) (Pl. 2, figs 17,18) has longer, narrower forewing, the costa is distinctly concave near apex, it has a small, dark discal spot and lacks metallic cilia; *Platytes cerussella* (Denis & Schiffermüller) (Pl. 2, figs 22,23) is much smaller and the cross-lines, of which two are usually visible, are strongly notched near dorsum.

Single-brooded; flies in June and July. The moth is easily disturbed from among tall grasses by day and comes freely to light.

LARVA Feeds through late summer and autumn until October on culms of various grasses, without forming silken galleries. Hibernates in a cocoon at ground level, in which pupation occurs in spring. Can sometimes be a pest in upland pasture: in an outbreak in Yorkshire in 1941, up to 128 cocoons per square foot of ground were reported (Beirne, 1952).

DISTRIBUTION Throughout Britain to the Shetlands, Outer Hebrides, Isles of Scilly, and in Ireland; also the Channel Islands. Common to abundant on the mainland of Britain and Ireland. Has been reported at lightships well out to sea and evidently migrates at times.

Crambus pascuella (Linnaeus)

Pl.1, figs 10(♂), 11(♀)

IMAGO Wingspan 22–25mm. Forewing with triangularly produced apex; light brownish ochreous, weakly pale-streaked towards dorsum; fine, dark subterminal line angled near costa; a fine, whitish longitudinal streak extends along dorsum to near base of subterminal line; a broad, shining white median longitudinal streak which touches basal one-third of costa and narrows to subterminal line, which it reaches but does not exceed; shortly before this point it is cut by a very oblique, fine dark line; near apex there is a small, narrowly triangular white costal spot which precedes the subterminal line, and a triangular white mark in apex itself. Hindwing smoky white. Cilia of both wings silvery metallic. A melanic form, ab. *obscurellus* Kuchlein, occurs rarely in the bogs of the New Forest, Hampshire, and in Kent.

There are five native *Crambus* species in which the longitudinal median streak is cut by a fine dark line: in *C. silvella* (Hübner) (Pl.1, fig.13), the ground-colour is a little darker and more conspicuously streaked, the white median streak barely touches the costa, and only at the base, and extends as a short white patch beyond the subterminal line; the dark line which cuts the streak is less oblique. *C. uliginosellus* Zeller (Pl.1, fig.14) is smaller and paler with relatively broader forewing, the white longitudinal streak touches the costa to half-way and is notched on the dorsal side opposite the point where it separates from the costa; the median streak reaches the termen but is obscured beyond the subterminal line through being bisected by a dark vein; the hindwing is whiter. *C. ericella* (Hübner) (Pl.1, fig.15) is darker, more uniform chocolate-brown, and the median longitudinal streak is narrower, separated from the costa throughout its length by a band of ground-colour at least as wide as the streak. In *C. pratella* (Linnaeus) (Pl.1, fig.16) and *C. lathoniellus* Zincken (Pl.1, figs 17–19), the median longitudinal streak is narrowed towards the base of the wing and is preceded anteriorly by a narrow white costal streak, and in both these species the hindwing is dark. Differences between *C. pascuella* and the vagrant *C. leucoschalis* Hampson (Pl.1, fig.12) are discussed under that species, *q.v.*

Single-brooded; flies from June to August. The moth is readily disturbed by day from among grasses and bushes in grassy places, and comes freely to light.

LARVA The early stages are apparently undescribed!

DISTRIBUTION Throughout the British Isles including the Channel Islands. Common to very common in pastures, grassy woodland rides, marshes and bogs.

Crambus leucoschalis Hampson

Pl.1, fig.12

IMAGO Similar to *C. pascuella* (Pl.1, figs 10,11) but rather larger. Distinguished from that species by the presence of a narrow white line running the whole length of

25

the costa, by the absence of a dividing line through the median longitudinal streak, by the presence of a dorsal notch or hook half-way along the median streak, and by the broadly white dorsum (Huggins, 1959).

One specimen taken in 1920 at Plymouth by Eng.-Capt. S. T. Stidson. A native of the drier parts of southern Africa, it could have been a genuine immigrant or introduced via the seaport: the latter seems more likely.

Crambus silvella (Hübner)

Pl. 1, fig. 13

IMAGO Wingspan 22–26mm. Forewing apex squarer than in *C. pascuella* (Pl. 1, figs 10,11). Forewing light brownish ochreous, rather heavily dark-streaked along veins, those dorsal to the outer one-quarter of the median longitudinal streak usually metallic scaled; subterminal line very fine, pale-edged; dorsum with more or less conspicuous pale streak; median longitudinal streak narrowly blackish-edged, touching costa only at very base and extending as a pale wedge beyond subterminal line to cilia, broken by a slightly oblique dark line two-thirds of wing length from base; whitish apical markings narrower and less conspicuous than in *C. pascuella*. Hindwing whitish fuscous. Cilia of all wings silvery metallic. Most like *C. pascuella*, and distinguished by the median streak extending beyond subterminal line: for other differences, see under *C. pascuella*.

Single-brooded; July and August. Seldom seen by day, but it becomes active just before dusk and flies well into the night, when it comes to light.

LARVA Feeds on sedges (*Carex* spp.) growing on boggy moorland, in a spinning, and pupates in a cocoon in the soil.

DISTRIBUTION A local species which occurs on boggy heathland in Surrey, Hampshire and Dorset, Norfolk and south-west Ireland.

Crambus uliginosellus Zeller

Pl. 1, fig. 14

IMAGO Wingspan 18–23mm. Forewing shorter and broader than in other *Crambus* species, the costa more arched, apex weakly produced; whitish or yellowish ochreous, with sparse scattering of rust-coloured scales towards dorsum; subterminal line white, edged on each side by rust-coloured scales; median longitudinal streak edged dark rusty brown, touching costa to half-way from base of wing, weakly hooked dorsally opposite this point, cut by an oblique dark line and extending conspicuously to termen beyond subterminal line; apical region with small white costal triangles preceding and beyond subterminal line. Hindwing white, often weakly suffused grey towards apex. Cilia of all wings metallic. Looks whiter than other *Crambus* in flight, and distinguished by broad white median longitudinal streak which touches costa to half-way.

Single-brooded; flies in June and July. Readily disturbed by day, and flies naturally towards dusk and after dark.

LARVA Early stages unknown.

DISTRIBUTION Very local in wet bogs in England and Wales, associated particularly with patches of deer-grass (*Scirpus cespitosus*). Also recorded from the Isle of Man (Chalmers-Hunt, 1970), Easterness (Agassiz, 1983), and Isle of Rhum (Wormell, 1983).

Crambus ericella (Hübner)

Pl.1, fig.15

IMAGO Wingspan 21–28mm. Apex of forewing very weakly produced; forewing greyish brown with golden gloss; subterminal line angulated, whitish interspersed with darker scales, edged darker on each side; a clear, tapering whitish streak along dorsum; median longitudinal streak narrow, slightly narrower at base, broken by oblique dividing line and terminated sharply at subterminal line; a small white costal mark just preceding subterminal line, and a clear small white triangle in apex; cilia with dark band, only weakly metallic. Hindwing fuscous, with white metallic cilia. More strongly and simply marked than other species of *Crambus*, and most like *C. pratella* (Linnaeus) (Pl.1, fig.16) which has, however, a rather strongly produced apical triangle and median longitudinal streak with conspicuous notch or hook half-way along dorsal margin.

Single-brooded; July, extending into August. Easily flushed by day, especially during the afternoon. Comes to light.

LARVA Early stages unknown.

DISTRIBUTION Local on northern moorland from Perthshire to the Orkneys and Outer Hebrides; isolated colonies in Cumberland (Vine Hall, 1959) and Yorkshire (Pelham-Clinton, pers. comm.) on limestone crags.

Crambus hamella (Thunberg)

Pl.1, fig.20

IMAGO Wingspan 24–27mm. Forewing slightly produced at apex; grey-brown, sometimes with a distinct mauvish tint, and with golden russet gloss; subterminal line distinct, angulated, whitish with metallic scales, edged dark reddish brown; median longitudinal streak broad, white, not touching costa, terminating in a point short of subterminal line, with a distinct dorsal notch slightly more than half-way from base, unbroken anterior half of apex containing a dark purplish brown triangular mark, behind which is a smaller white triangle; termen containing four or five small intense black dots. Hindwing light fuscous. Cilia of both wings silvery metallic.

The characteristically shaped longitudinal stria precludes confusion with any other species.

Single-brooded; in the second half of July and well on into late August. Usually sluggish and difficult to disturb by day, but occasionally as active as any of the others of the genus. Flies at dusk, comes to light, and may be found at heather bloom and resting on heath vegetation at night.

LARVA Early stages unknown, though the moth appears to be associated with grasses of dry heathland such as wavy hair-grass (*Deschampsia flexuosa*).

DISTRIBUTION Locally common on dry heaths from southern England to south Scotland, and recorded from Killarney, Co. Kerry, in south-west Ireland (Beirne, 1952); also the Channel Islands.

Crambus pratella (Linnaeus)
= *dumetella* (Hübner)

Pl.1, fig.16

IMAGO Wingspan 22–25mm. Forewing with apex rather strongly produced, warm light brown with irroration of silvery scales, especially along distal region of veins; subterminal line angulated, silvery metallic with dark brown edging towards base; a fine white subcostal line extends nearly half-way along wing; median longitudinal streak broadens from narrow base, anterior edge angulated dorsally beyond middle, posterior edge notched at half-way, reaching subterminal line but not extending beyond it, and broken half-way between anterior angulation and subterminal line by an oblique dark line; small white costal marks are present on either side of subterminal line, and apex contains a small white triangle. Hindwing greyish fuscous. All cilia whitish metallic.

C. *pratella* can be confused only with the rather smaller C. *lathoniellus* (Pl.1, figs 17,19), in which the median longitudinal streak is narrower with the anterior edge straight, not angulated.

Single-brooded; flies in June and July. Though chiefly nocturnal, it may sometimes be found in numbers flying in evening sunshine.

LARVA July to May on roots and stem bases of grasses growing in dry, sandy places, inhabiting a silken tube to which frass and plant debris are attached. Pupation occurs in a cocoon at the end of the larval tube.

DISTRIBUTION One of the more local species of Crambinae, on dry, grassy pastures where the turf is short; from southern England at least as far north as Inverness, in several localities in Ireland, and in Guernsey, Jersey and Alderney.

Crambus lathoniellus Zincken
= *nemorella* auctt.

Pl.1, figs 17,19(♂♂),18(♀)

IMAGO Wingspan 18–22mm. Forewing moderately produced at apex; in male, brown with greyish interneural streaks and more or less heavy irroration of blackish scales; in female, wing more ochreous with whiter interneural streaks and less irroration, making it look considerably paler; subterminal line has as its most conspicuous feature a dark, angulated inner edge, beyond which is a patch composed of shining white scales; fine whitish or yellowish subcostal streak which occupies basal half of wing is sometimes obliterated; median longitudinal streak narrow, widening from base, anterior edge straight, dorsal edge notched, extending to subterminal line and interrupted a short way before it by an oblique dark line. Hindwing brownish fuscous. Cilia of both wings whitish, moderately metallic.

A rather variable species: some very dark specimens occur in the bogs of western Ireland (fig. 19). The only species to be confused with *C. lathoniellus* is the far less common *C. pratella* (Pl. 1, fig. 16), *q.v.*

The first crambid to appear on the wing in Britain, flying in an apparently extended emergence from May to August. It rests head downwards on tall grasses by day, but is very lively and alert. Its natural flight-time is probably at night, when it comes to light.

LARVA On roots of many species of grass, living from late summer till April in a tubular silken gallery amongst grass roots. Pupation occurs in a firm silken cocoon among the roots or attached to a stone.

DISTRIBUTION Occurs almost throughout the British Isles, usually in great profusion; apparently uncommon in Orkney and Shetland.

Crambus perlella (Scopoli)

Pl. 1, figs 21, 22

IMAGO Wingspan 22–29mm. Apex of forewing bluntly angled without being produced. In the typical form, forewing bright, shining white or ochreous white, without markings; in f. *warringtonellus* Stainton (fig. 22), veins and, to a varying extent, the ground-colour are suffused dusky grey, producing a heavily streaked effect. Hindwing whitish, more or less suffused grey. Cilia whitish, glossy without being metallic. Both forms are easily recognised. See also *Catoptria lythargyrella* (Hübner) (Pl. 2, fig. 11).

Single-brooded; flies in July and August. It rests by day head downwards on tall grasses and is very readily disturbed; it flies at night and comes to light; it has been recorded at lightships, evidence that it migrates at times.

LARVA Feeds on bases of stems of grasses from a tubular silken gallery which sometimes extends upwards a few centimetres from ground level; September to June. Pupation occurs in a cocoon at or just below ground level.

DISTRIBUTION Throughout the British Isles and the Channel Islands, in all kinds of grassy habitats, and often abundant. The f. *warringtonellus* occurs in all populations, but predominates in boggy areas and in the north and west.

Agriphila selasella (Hübner)

Pl. 1, fig. 29

IMAGO Wingspan 22–30mm. Face without projecting cone, only slightly prominent. Forewing bluntly squared at apex; sandy whitish yellow or ochreous, more or less suffused fuscous anterior to median longitudinal streak; streak c.1mm broad at widest point, terminating in up to four fine branches which extend along veins; cilia shining whitish or fuscous. Hindwing grey.

The species can be confused only with certain forms of *A. tristella* (Denis & Schiffermüller) (Pl.1, figs 26,27): that species has a distinct facial cone, besides having traces of a subterminal line and yellow or creamy, never white, median longitudinal streak; termen more rounded, less square than in *A. selasella*. The forewing of *A. selasella* has a distinctly smoother texture than that of *A. tristella*.

Single-brooded; flies in July and August. Can be disturbed by day from amongst grasses, but the usual flight-time is at night, when it will come to light. It may also be found sitting around on grass stems after dark. It has been captured at lightships off the British coast.

LARVA Inhabits a silken gallery covered with grass debris, and feeds on grasses including common saltmarsh-grass (*Puccinellia maritima*), small cord-grass (*Spartina maritima*) and sheep's-fescue (*Festuca ovina*): it is full-fed in June.

DISTRIBUTION Local in England, Wales, southern Scotland (Pelham-Clinton, pers. comm.), Ireland and the Channel Islands; a characteristic species of saltmarshes, but also found in fens and freshwater marshes, and less frequently along wood borders.

Agriphila straminella (Denis & Schiffermüller)
= *culmella* auctt.

Pl.1, figs 23(♂), 24,25(♀♀)

IMAGO Wingspan 16–20mm. Forewing weakly pointed at apex, narrower in female; whitish straw-coloured, suffused brownish fuscous between usually inconspicuous branched median longitudinal streak and costa, and along termen; sometimes the dark suffusion is heavier and extends to dorsal region, when median streak shows as a fine pale line from base, which branches from the cell along three or four veins to termen. Uniformly brown specimens occur locally from the Shetlands (fig.25) and Orkneys (Lorimer, 1983) to Surrey (in British Museum (Natural History) coll.). Hindwing grey, normally darker than forewing. Cilia of both wings shining white metallic. Unmistakable; but see also the larger *Chrysoteuchia culmella* (Pl.1, figs 7,8).

Single-brooded; flies from June to August. Readily disturbed by day and also flies naturally both by day and at night, when it is an abundant visitor to light.

LARVA September to June on grasses, especially smaller species such as sheep's-fescue (*Festuca ovina*) in galleries amongst the foodplants. Pupation occurs in a silken frass-covered cocoon.

DISTRIBUTION Abundant throughout the British Isles in all kinds of grassy habitat.

Agriphila tristella (Denis & Schiffermüller)
Pl.1, figs 26(♂), 27(♀); *frontispiece*, fig. 1

IMAGO Wingspan 25–30mm. Forewing pale whitish straw-coloured to rich tawny yellow or sometimes dark brown, more or less suffused fuscous especially towards costa anterior to pale creamy or yellow, branched median longitudinal streak; cross-lines rather indistinct, often obsolete, antemedian extending obliquely basad to dorsum, subterminal weakly bowed basad and often represented by disjunct patches of dark scales over the veins; fringe whitish with grey margin and narrow subbasal line. Hindwing grey, fringe white with narrow dark band very close to base.

Usually more tawny and rougher-textured than *A. selasella* (Pl.1, fig.29), the only other species with which *A. tristella* could be confused; other differences are given under the description of that species.

Single-brooded; flies from July to September. Readily disturbed by day from amongst tall grasses, and freely attracted to light. Known to migrate.

LARVA September to June in silken galleries amongst the bases of grass-stems. Pupa in an oval, silken, frass-covered cocoon in soil amongst grass roots.

DISTRIBUTION Common to abundant throughout the British Isles, with the exception of the Shetlands, in places where tall grasses abound. Recorded from the Channel Islands.

Agriphila inquinatella (Denis & Schiffermüller)

Pl.1, figs 30,32(♂♂), 31(♀)

IMAGO Wingspan 23–29 mm. Apex of forewing rather square and sharply angled; forewing whitish to pale straw-coloured, variably suffused fuscous along veins, and occasionally over the whole wing with the exception of a fine pale median longitudinal streak; a pale form occurring on Dungeness beach has the suffusions ochreous; an oblique median cross-line passes through a short dark longitudinal dash lying within the cell; the curved subterminal line is often thickened to form a chevron-like mark slightly more than half-way to dorsum; a dark basal streak which extends to near the median cross-line is usually discernible; termen with a row of small black dots; fringe pale grey, outer half darker grey. Hindwing grey, fuscous-tinted, more so in dark specimens; cilia light grey. A variable species, but easily separable from all but *A. geniculea* (Haworth) (Pl.2, figs 1,2), *Pediasia contaminella* (Hübner) (Pl.2, figs 17,18) and *P. aridella* (Thunberg) (Pl.2, fig.19). *A. geniculea* has a similar dark dash in the cell, though slightly more oblique, but is most easily distinguished by the cross-lines, both of which are strongly elbowed; in *P. contaminella*, the costa is weakly concave towards apex and the cell-mark is small, round and dot-like; *P. aridella* is also a variable species, in it the cell-mark is usually indistinct and merged with a dark median longitudinal streak, anterior to which is a less conspicuous paler streak; the subterminal chevron-like mark is less oblique than in *A. inquinatella* and more blurred, and there is a small dark dot near the tornus of the hindwing, variable in expression but seldom entirely absent.

Single-brooded; flies from July to September. Readily disturbed by day from short grass, on the stems of which it rests head downwards, or on the ground. Flies from dusk onwards, comes to light and occasionally to sugar.

LARVA Feeds in a slight silken gallery amongst roots and stem-bases of smaller grasses, especially sheep's-fescue (*Festuca ovina*). Pupates in May.

DISTRIBUTION Widespread but local in mainland Britain south of Perthshire, commoner in the south and east; Inner Hebrides (Wormell, 1983); Isles of Scilly, common (Agassiz, 1981b). In Ireland on the east coast and in the south-west. The Channel Islands. Associated with smaller grasses on light, dry soils, grassy shingle, sandy heathland and dunes, calcareous fields and downs.

Agriphila latistria (Haworth)

Pl.1, fig.28

IMAGO Wingspan 22–27mm. Forewing ferruginous brown with greyish tinge, unmarked except for broad white median longitudinal stripe which extends right into the fringe, and row of minute black dots along termen; fringe grey, paler in distal half, with exception of white area. Hindwing whitish grey with clear white cilia. Unmistakable: in no other British species does the median streak extend on to the fringe.

Single-brooded; flies in July and August. Lies hidden by day amongst heather and grasses; flies gently at night and may be found resting on stems of heather and grass, often *in copula*.

LARVA Tunnels close to soil surface amongst roots of grasses, especially *Bromus* spp. (Beirne, 1952). Pupates in a long, rather narrow silken sand-covered cocoon.

DISTRIBUTION Local but fairly common where it occurs, on coastal sandhills, dry heaths and, more rarely, borders of woodland rides in sandy districts. Recorded from the coasts of England from Lincolnshire and Lancashire southwards, on southern heaths such as those of the New Forest, Hampshire, and from Monmouthshire, Ayrshire, Perthshire and the Isle of Arran. In Ireland, from Co. Cork (Agassiz, pers. comm.). One record from Jersey (Long, 1967).

Agriphila poliellus (Treitschke)

Pl.1, fig.33

IMAGO Wingspan 19–26mm. Female smaller but heavier-bodied than male, with more pointed forewing. Forewing narrow, light brownish grey, paler in median area, irrorate dark fuscous; a small, fuscous discal dot above pale median streak, and a series of small, black terminal dots; fringe metallic greyish. Hindwing light grey with paler fringe.

One specimen taken at Deal, Kent, 1885 is in British Museum (Natural History). Local in central and eastern Europe in dry grassland.

Agriphila geniculea (Haworth)

Pl.2, figs 1, 2

IMAGO Wingspan 20–26mm. Forewing whitish, heavily irrorate ashy grey brown, with indistinct paler median longitudinal streak; median cross-line oblique, elbowed, thickened and darkened in cell to form an oblique median dash; subterminal line very strongly elbowed near apex, and again, but making a shorter arm, near tornus, the median part of this arm often thickened to form a chevron; a row of minute black dots along termen; fringe grey, silvery metallic. Hindwing light grey, fringe whitish grey. The strongly elbowed cross-lines distinguish this species from *A. inquinatella* (Pl.1, figs 30–32), *Pediasia contaminella* (Pl.2, figs 17,18) and *P. aridella* (Pl.2, fig.19) with which it might otherwise be confused.

Single-brooded; flies from July to October. Rests by day on the ground or on grasses, but particularly in bushes of young conifers, from which it is easily disturbed. Flies at night and comes to light.

LARVA September to June on grasses, living in a tubular silken gallery.

DISTRIBUTION Widespread and locally abundant in Britain and Ireland from Perthshire southwards, chiefly in dry pastures and particularly on coastal sandhills. Inner Hebrides (Wormell, 1983). Common in the Channel Islands.

Catoptria permutatella (Herrich-Schäffer)
= *myella* auctt. *nec* Hübner

Pl.2, fig.3; text figs 3a,b,g

IMAGO Wingspan 22–29mm. Forewing golden ferruginous ochreous, mahogany brown adjacent to broad dilating shining white median longitudinal stripe, which is bisected by an oblique mahogany coloured median cross-line and also traversed subterminally by another line which leaves a narrow white line distal to it, running nearly parallel to the termen; fringe greyish, banded with brown and weakly chequered opposite the longitudinal stripe. Hindwing whitish grey, slightly darker towards margin, fringe whitish. Distinguished from the widespread *C. pinella* (Linnaeus) (Pl.2, fig.6) by the white stripe slightly narrower at base, the more oblique median cross-line and particularly by the narrow white subterminal line which is absent in *C. pinella*. Can be separated from *C. osthelderi* (de Lattin), *q.v.*, and *C. speculalis* Hübner only by examination of genitalia, though in *C. speculalis* (Pl.2, fig.5) the subterminal white line curves away from termen as it approaches the tornus (Carter, 1967).

Single-brooded; July and August. Rests by day among the foliage of pine (*Pinus* spp.) and sometimes birch (*Betula* spp.), from which it can be disturbed by beating. It shows a preference for isolated trees. The natural time of flight is from dusk into the night.

LARVA Reported to feed on mosses (Bleszyński, 1965, referring to the genus).

DISTRIBUTION Reported from only a few glens in Perthshire and Aberdeenshire, notably Glen Tilt and Deeside as far east as Aberdeen itself.

Catoptria osthelderi (de Lattin)

Pl.2, fig.4; text figs 3c,d,h

IMAGO Indistinguishable on superficial characters from *C. permutatella*, but readily

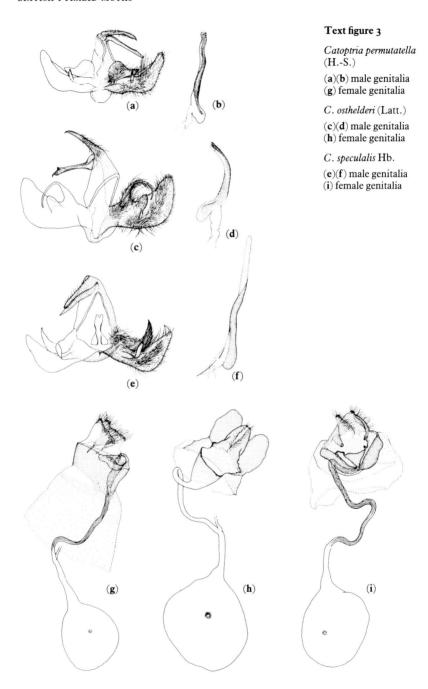

Text figure 3

Catoptria permutatella
(H.-S.)

(**a**)(**b**) male genitalia
(**g**) female genitalia

C. osthelderi (Latt.)

(**c**)(**d**) male genitalia
(**h**) female genitalia

C. speculalis Hb.

(**e**)(**f**) male genitalia
(**i**) female genitalia

separated from both that species and *C. speculalis* by differences in structure of the genitalia (see text fig. 3): in the male, the broader valvae which bear on the costa a long, strongly curved spine, and the dorsal protuberance on the gnathos serve to separate it from both the others; in the female, *C. ostheldeeri* possesses a heavily sclerotized, dorsally expanded and deeply notched ostial plate which in *C. permutatella* is a simple, collar-like structure and in *C. speculalis* has a strongly sclerotized ventral lip (Carter, 1967).

DISTRIBUTION One female taken at a blended mercury-vapour/tungsten light at Bexley Heath, Kent, 29 July 1962 (Roche, 1963). It has a rather eastern distribution in Europe, but occurs in the Netherlands, Denmark and Germany.

Catoptria speculalis Hübner

Pl.2, fig.5; text figs 3e,f,i

IMAGO Very similar to the two preceding species; forewing with subterminal white line which curves away from termen towards tornus, not running parallel with termen as in the other two species. The genitalia are quite different in both sexes (Carter, 1967).

DISTRIBUTION One specimen in the British Museum (Natural History) labelled 'W. Reid, Perth, July 1890' (Carter, *loc.cit.*). In Europe, an alpine species.

Catoptria pinella (Linnaeus)

Pl.2, fig.6; *frontispiece*, fig. 2

IMAGO Wingspan 20–26mm. Forewing golden ferruginous ochreous with broad, dilating, shining white median longitudinal stripe, which is bisected by an oblique mahogany brown median cross-line and edged distally by similar colour; subterminal line distinct only between apex of white stripe and costa; fringe grey with dark brown line, weakly metallic anteriorly. Hindwing whitish grey; fringe white with darker line. Distinguished from the three preceding species by the absence of fine white subterminal line at tip of white stripe.

Single-brooded; flies in July and August, sometimes September. Conceals itself by day amongst foliage of trees and bushes, especially pine, and is not often seen. Flies at night and comes to light.

LARVA Lives in a small vertical silken tube amongst dense tufts of cottongrass (*Eriophorum* spp.), tufted hair-grass (*Deschampsia caespitosa*) and other grasses, from September to June. Pupa in a white silken cocoon covered with grass particles, in a leaf sheath of the foodplant.

DISTRIBUTION Widespread but local and generally rather uncommon, chiefly in heathy woodlands and on boggy ground. Most frequent in the south and east of England but extending to Perthshire, Aberdeenshire and the Inner Hebrides (Pelham-Clinton, pers. comm.) and to Ireland. Also in the Channel Islands.

Catoptria margaritella (Denis & Schiffermüller)

Pl.2, fig.7

IMAGO Wingspan 20–24mm. Forewing light ferruginous brown, paler towards dorsum, with a broad, dilated, shining white median longitudinal stripe, broadest at four-fifths then tapering towards apex and ending just short of termen; fringe grey with indication of darker line, weakly metallic. Hindwing grey, fringe white. Readily separable from *C. permutatella* and its close allies (Pl.2, figs 3–5) by the absence of a cross-line dividing the median longitudinal stripe; *C. furcatellus* (Zetterstedt) (Pl.2, fig.8) has uniformly coloured, mahogany-brown forewing with narrower, dull white median longitudinal stripe.

Single-brooded; flies in July and August. Easily disturbed by day from amongst bog and moorland vegetation; flies at night and comes to light; after dark, the sluggish females may be found resting on grass-stems, often *in copula*.

LARVA Undescribed.

DISTRIBUTION Locally common on boggy heaths and moors in Scotland including Orkney and the Hebrides, the mosses of northern England as far south as Staffordshire and north Norfolk, Cornwall (Bodmin Moor), Devon, Somerset and parts of central and north Wales. Widespread but local in Ireland.

Catoptria furcatellus (Zetterstedt)

Pl.2, fig.8

IMAGO Wingspan 19–23mm. Forewing uniformly mahogany brown or dark ferruginous fuscous with rather narrow, dull white median longitudinal stripe, hardly dilated, broadest at four-fifths, then tapering to a little short of termen; cilia white, brown-based, shining. Hindwing light to dark grey, often with paler longitudinal wedge-shaped area; cilia very pale grey, white based and thus forming a clear white basal band on fringe. *C. margaritella* (Pl.2, fig.7) is the only other *Catoptria* species with an unbroken median longitudinal stripe: differences are given under that species.

Single-brooded; July and August. In calm, warm weather it makes short flights by day over the sparse turf of its habitat; both sexes are to be encountered at these times.

LARVA Undescribed.

DISTRIBUTION Extremely local on mountains between 400m and 900m in north Wales, the Lake District and the Scottish Highlands, and recorded from Isle of Rhum (Wormell, 1983). Common at lower elevation on Ronas Hill, Shetland.

Catoptria falsella (Denis & Schiffermüller)

Pl.2, fig.9

IMAGO Wingspan 18–24mm. Forewing whitish, irrorate blackish brown between the veins; a dilated white median longitudinal stripe broadly and obliquely divided by a light brown cross-line which is heavily irrorate fuscous between the veins; subterminal line strongly bent near apex and double-angled at tornus; fringe irregularly chequered and lined dark brown and white, weakly metallic. Hindwing whitish grey, fringe shining white with darker line near base. Unmistakable: *C. verellus* (Zincken) (Pl.2, fig.10) lacks the white median longitudinal stripe.

Single-brooded; flies in July and August. Hides by day amongst moss on walls or in thatch or ricks or in dense evergreen foliage and is seldom encountered. Flies at night and comes to light.

LARVA Feeds till May in a silk-lined gallery amongst mosses on old walls, feeding on *Tortula* and *Barbula* spp. and perhaps other mosses growing there. Pupa in larval gallery.

DISTRIBUTION Local and rather uncommon from Aberdeen southwards; most frequent in southern England in villages where thatch and mossy walls abound. Also in the Channel Islands.

Catoptria verellus (Zincken)

Pl.2, fig.10

IMAGO Wingspan 17–18mm. Forewing brown suffused blackish except towards costa and on veins; costa whitish towards base, and a weak, whitish median basal streak which soon becomes obsolete; second line fine, white, indented near dorsum then curved towards termen above middle, where it becomes unclear before bending sharply basad to meet costa very obliquely; basal to this line and immediately above the tornal indentation is a narrow white wedge-shaped mark; fringe grey, somewhat metallic, finely barred. Hindwing light grey with paler fringe. 'The scruffiest in appearance of all our Crambids', and easily overlooked as a small, dusky *C. falsella* (Pl.2, fig.9), or ignored altogether (Huggins, 1954).

Flies in July and August; comes to light.

LARVA Amongst moss on branches of old apple, plum and poplar trees; in spring until May (Bleszyński, 1965).

DISTRIBUTION Very little is known about the status of this species in Britain. Four specimens taken at light at Haslingfield, Suffolk, by A. F. Griffith in August, 1877 (one) and July 1878 (three) (Griffith, 1881) might suggest that it is a scarce and obscure resident, though all the other records, from Folkestone, Kent and Bognor, Sussex, suggest immigration. It is widespread on adjacent parts of the Continent.

Catoptria lythargyrella (Hübner)

Pl.2, fig.11

IMAGO Wingspan 24–34mm. Forewing clear, glossy light ochreous yellow, some-times greyish-tinged, the veins very slightly paler but otherwise unmarked; cilia irregularly mixed whitish and ochreous. Hindwing whitish grey, fringe white. A species that could presumably be overlooked amongst other common Crambinae, but the completely unmarked, ochreous forewing will readily distinguish it from such as *Agriphila selasella* and *A. tristella*. It might be mistaken for one of the darker forms of *Crambus perlella*, but is larger and less glossy. (See Pl.1, fig.22.)

Five specimens were taken at Deal, Kent, in 1889 and are in the British Museum (Natural History); it has not been recorded again in this country. Native of central Europe and Asia Minor.

Chrysocrambus linetella (Fabricius)
= *cassentiniellus* (Herrich-Schäffer)

Pl.2, fig.12; text figs 4a,b,e

IMAGO Wingspan 20–27mm. Forewing broad, cream-coloured to yellowish with golden brown or rusty brown stripes between the veins, the two in the apex not coalescent; median cross-line thick, almost transverse, weakly angled or sinuate; second line finer, obtusely angled at middle, both lines rusty brown; termen with series of small, black, interneural dots; fringe golden metallic. Hindwing brownish, in female paler towards base; fringe white.

Chrysocrambus craterella (Scopoli)
= *rorella* (Linnaeus)

Pl.2, fig.13; text figs 4c,d,f

IMAGO Wingspan 19–22mm. Very similar indeed to *C. linetella* (Pl.2, fig.12) though with the two rusty brown stripes in apex of forewing thickened and coalescent.

Species of *Chrysocrambus* can only be separated with certainty by examination of the genitalia:

	C. linetella	*C. craterella*
Male	(text figs. a,b)	(text figs. 4c,d)
gnathos	apex dilated, squared at tip	apex not dilated, rounded at tip
valve	about twice as long as broad, cucullus tapered, costal arm absent	about five times as long as broad, cucullus parallel sided, costal arm broadly tapered
aedeagus	small, lacking cornuti	very large, with two well-developed cornuti
Female	(text fig. 4e)	(text fig. 4f)
ostium	ductus strongly sclerotized	ductus usually weakly sclerotized
bursa copulatrix	signum minute, hardly discernible	two larger signa, weakly developed

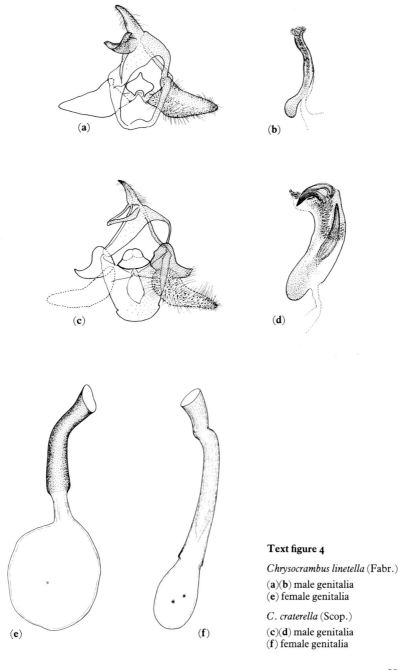

Text figure 4

Chrysocrambus linetella (Fabr.)
(a)(b) male genitalia
(e) female genitalia

C. craterella (Scop.)
(c)(d) male genitalia
(f) female genitalia

The genus is widespread in the Mediterranean region, and two other species, *C. dentuellus* (Pierce & Metcalfe) and *C. sardiniellus* (Turati) (= *cornutellus* sensu Whalley, 1959) could appear in this country. Photographs and genitalia figures are given by Whalley (1959).

Specimens of both *C. linetella* and *C. craterella* from Kent, and of *C. craterella* from Boxhill, Surrey, are in British Museum (Natural History) (Whalley, *loc. cit.*). Both are assumed to be rare immigrants, but could be overlooked amongst populations of *Thisanotia chrysonuchella* (Scopoli) (Pl.2, fig.14), *q.v.*

Thisanotia chrysonuchella (Scopoli)

Pl.2, fig.14

IMAGO Wingspan 23–26mm. Forewing broad as in species of *Chrysocrambus*, whitish with brown shading in a broad band along costa and in two narrow bands between cell and dorsum, and distally between the veins, the whole wing lightly irrorate with large black scales; median cross-line oblique, weakly elbowed, brown and merging with longitudinal bands of shading; subterminal line strongly curved anteriorly, thin, brown, edged white distally; fringe dark grey, paler basal region with fine dark line, golden metallic. Hindwing brownish grey with fine dark margin; fringe whitish with fine dark line near base, glossy. An unmistakable species: the *Chrysocrambus* species have wings of similar shape but with fine dark longitudinal striae and sharply defined, narrow median and subterminal cross-lines.

Single-brooded; flies in May and June. Easily disturbed by day, especially in the afternoon, from among tall grasses on which it rests. Flies from dusk onwards and comes to light.

LARVA Feeds from July to May on stem bases of sheep's-fescue (*Festuca ovina*) and other grasses, living in a silken tube. Pupa in a cocoon in the larval tube.

DISTRIBUTION Local on southern chalk downs, coastal sandhills and cliffs, and in Breckland, usually rather common where it occurs, but generally one of the less frequently encountered diurnal species.

Pediasia fascelinella (Hübner)

Pl.2, figs 15(♂),16(♀)

IMAGO Wingspan 24–30mm. Forewing pale whitish ochreous, veins paler, weakly irrorate fuscous between cell and dorsum forming weak, ill-defined dark streaks; median and subterminal cross-lines parallel, oblique, curved basad towards costa, composed of series of brown or fuscous spots separated by the pale veins; fringe strongly chequered white and brown. Hindwing brownish white, fringe white. Though variable in the amount of fuscous irroration, this is an unmistakable species: the chequered fringe distinguishes it at once.

Single-brooded; flies in July. Very sluggish by day, hiding in crannies and amongst the stem bases of marram grass, but active from dusk onwards, when it flutters up grass-stems, and flies close to the ground.

LARVA Feeds from September to June on root-stocks of sandhill grasses, inhabiting a long tube of silk and sand particles, in the hind part of which frass accumulates. Pupa in a vertical cocoon of silk and sand outside the mouth of the larval gallery.

DISTRIBUTION Extremely local on sandhills on the coasts of Lincolnshire, Norfolk, Suffolk, Essex and south Devon, in the area of stabilizing sand behind the main dunes. Recorded from Jersey (Long, 1967).

Pediasia contaminella (Hübner)

Pl.2, figs 17,18

IMAGO Wingspan 20–30mm. Forewing dull greyish ochreous or reddish ochreous, irrorate to a greater or lesser extent with minute fuscous specks; a small dusky point in the discal region; cross-lines very faint, the median oblique, angled on to costa, the subterminal toothed at each vein and strongly double-angled in tornus; fringe dark grey with narrow pale base. Hindwing light grey, fringe slightly paler. A melanic form, ab. *sticheli* Constant (fig.18) occurs in which the forewings are heavily irrorate blackish fuscous. Once known, a readily recognised species: alive, its resting posture, head down and abdomen obliquely raised, is characteristic; the forewing is duller-coloured than in similar species, the discal spot is a point, not a short bar as in *A. inquinatella* (Pl.1, figs 30–32) and *A. geniculea* (Pl.2, figs 1,2), and the wing is less glossy and lacks the fuscous streaking of *P. aridella* (Pl.2, figs 19,20); in none of these species is the subterminal line toothed. See also *Chrysoteuchia culmella* (Pl.1, figs 7,8).

Flies in July and early August; occasionally there is a small emergence in late September, as in 1976. Rests concealed by day and flies from dusk onwards, coming readily to light.

LARVA Feeds from September to May on grasses such as sheep's-fescue (*Festuca ovina*) in dry localities, inhabiting a funnel-shaped silken tube placed vertically among the grass tufts. Pupa in a tough, silken, sand-covered cocoon.

DISTRIBUTION Local in dry grassland, on golf courses and playing fields in Essex, Kent, Middlesex and Hertfordshire and extending at least as far as east Hampshire, possibly more widespread and undoubtedly overlooked. Isles of Scilly (Agassiz, 1981b). The Channel Islands.

Pediasia aridella (Thunberg)

= *salinellus* (Tutt)

Pl.2, figs.19,20

IMAGO Wingspan 20–26mm. Forewing ochreous, rather paler on veins, slightly glossy; a median longitudinal streak of blackish suffusion from base to about half-way, where it meets a more or less well-defined, oblique, blackish fuscous cross-line; subterminal line oblique, brown, elbowed towards costa and with a chevron-shaped mark of fuscous suffusion a little over half-way to tornus; several black specks on margin of termen towards tornus; fringe concolorous, slightly glossy. Hindwing light grey, with small dark tornal spot or short bar; fringe white, glossy, with faint darker basal line. Glossier-winged than *P. contaminella* (Pl.2, figs 17,18), which lacks the fuscous streaking; this species lacks the bar-shaped discal mark characteristic of *Agriphila inquinatella* (Pl.1, figs 30–32), in which the hindwing is devoid of marking.

Single-brooded; flies from June to August. Hides by day and is seldom encountered until after dark when it flies close to the ground, rests upon the stems of saltmarsh grasses, and comes to light.

LARVA Feeds from September to May on stem bases of common saltmarsh grass (*Puccinellia maritima*) and perhaps other species of grass which grow in the drier parts of salterns, living in a tubular silken frass-covered gallery on the ground. Pupa in an oval cocoon covered with particles of soil.

DISTRIBUTION Local along the dry margins of salterns from Spurn Head, Yorkshire to the Isle of Wight; however, the distribution is imperfectly known and the species is possibly more widespread.

Platytes alpinella (Hübner)

Pl.2, fig.21

IMAGO Wingspan 18–22mm. Apex of forewing with a characteristic and conspicuous triangular projection. Forewing whitish irrorate fuscous along dorsum and with broad fuscous band along costa which extends into apical projection; dorsal to this band is a fine white median longitudinal stripe most clearly defined in basal two-thirds of wing; median cross-line ill-defined, oblique, fine, angulated; subterminal line fine, oblique, angled basad towards costa and distad into tornus, brown, white edged; cilia of unequal length, whitish, shining. Hindwing grey, fringe paler with dark line at base. The characteristic shape of the forewing, together with dark costal band and fine median longitudinal stripe prevent confusion with any other species.

Single-brooded; flies in July and August. Remains hidden by day and is seldom disturbed; flies at dusk and after dark, comes to flowers of ragwort and to light.

LARVA Stated to feed on the moss *Tortula ruraliformis* (Bleszyński, 1965).

DISTRIBUTION Very local and rather uncommon on coastal sandhills and shingle in Yorkshire (Spurn Head), Lincolnshire, Norfolk, Suffolk, Essex, Kent, Sussex, Hampshire (Hayling Island), Isle of Wight, Devon and Isle of Man. Also recorded from Co. Cork, Ireland (Bond, 1979). Stragglers occasionally occur inland in the seaboard counties.

Platytes cerussella (Denis & Schiffermüller)

Pl.2, figs 22(♂), 23(♀)

IMAGO Wingspan of male 12–15mm, female 10–13mm. *Male.* Forewing ochreous brown or greyish brown, weakly irrorate ferruginous; median and subterminal cross-lines oblique, strongly angled basad towards costa and twice angled to dorsum, the subterminal more strongly; three small black dots on termen near tornus; fringe darker than ground-colour of wing, grey with darker line near base, metallic. Hindwing dark greyish brown, fringe whitish with ferruginous broken line near base. *Female.* Wings narrower than male. Forewing whitish irrorate ferruginous towards apex; cross-lines indistinct, terminal dots present, fringe whitish with ferruginous line near base. Hindwing paler than in male. The smallest crambid: general appearance of male not unlike *Chrysoteuchia culmella* (Pl.1, fig.8), though much smaller and with quite different cross-lines.

Single-brooded; flies in June and July. Rests by day amongst short grasses and sand sedge, from which the male is easily disturbed. The female is more sluggish, but will fly on very hot afternoons. The male comes to light, and the female can be found at night resting on short vegetation. There is another flight at sunrise.

LARVA Feeds in April and May on roots of small, stiff grasses growing on sand and shingle, perhaps also on sand sedge (*Carex arenaria*).

DISTRIBUTION Locally abundant on sandy or shingly coasts from Norfolk southwards, and inland on Breckland. Isles of Scilly (Agassiz, 1981b).

Ancylolomia tentaculella (Hübner)

Pl.2, fig.24

IMAGO Wingspan 30–34mm. Termen of forewing strongly sinuate. Forewing light ochreous, irrorate grey and with a few black scales, with a conspicuous narrow creamy white median longitudinal stripe which is weakly angled towards tornus beyond the cell; another fine, yellowish subdorsal longitudinal stripe and cream-coloured striations along costa and on veins in subterminal region; a small greyish ochreous discal spot is present; beyond the fine, curved, ochreous subterminal line, and occupying the whitish area between this and the termen is a zigzag brown line; silver metallic scales are present on this line and on some of the veins; fringe grey, bisected by a broad white line, highly metallic. Hindwing light grey with a fine dark line round termen; fringe white. Quite unlike any other British crambid.

Flies in June and July. On the Continent it is readily disturbed by day in grassy places, and comes to light.

LARVA Feeds from September to June, in a vertical tunnel 3–4cm long at the base of stems of large grasses, especially cock's-foot (*Dactylis glomerata*).

DISTRIBUTION Dungeness, Kent, one taken at dusk, 26 July 1935 (Morley, 1953); another, Dymchurch, Kent, at light, 5 July 1952 (Edwards & Wakely, 1952). There exists the remote possibility that the species is resident in the area, though it is more likely to have been an immigrant from southern Europe on both occasions.

Subfamily SCHOENOBIINAE (3 species)

The Schoenobiinae are long-winged, long-legged insects which are closely associated with waterside vegetation. The colour of their wings blends with the dead leaf-sheaths of reeds and sedges. They are sexually dimorphic, the females having much longer and more pointed wings than the males. The larvae feed within stems of reeds and sedges, and may float from plant to plant on rafts made from cut pieces of leaf.

Schoenobius gigantella (Denis & Schiffermüller)

Pl.2, figs 25, 27(♂♂),26(♀)

IMAGO Wingspan of male 25–30mm, female 41–46mm. Strongly sexually dimorphic. Labial palpi exceeding length of head and thorax combined. *Male.* Forewing rather broad, apex blunt, light ochreous fuscous, sometimes reddish-tinted or heavily irrorate blackish fuscous; two small, round dark discal spots usually present, and there are frequently other more or less conspicuous dots along the major veins and in the subterminal area. Hindwing greyish white, glossy, with postmedian and terminal bands of short, dark streaks. *Female.* Forewing relatively much narrower, with termen oblique and apex pointed, ochreous brown; commonly almost un-marked, but sometimes with blackish discal spot and other scattered spots, some forming an oblique subapical streak; more rarely there is heavy black suffusion in the anterior half of the wing from base to apex. Hindwing pure white. Male is unmistakable; female is considerably larger than that of *Chilo phragmitella* (Pl.1, fig.2), *Donacaula forficella* (Thunberg) (Pl.2, fig.29) and *D. mucronellus* (Denis & Schiffermüller) (Pl.2, fig.31); distinguished further from the first in that the costa near the apex is convex, not concave; the female of *D. forficella* has apex of forewing even more pointed, with termen distinctly concave; the forewing of *D. mucronellus* has a broad, pale costal stripe sharply demarcated from a dark subcostal streak from base to apex.

Flies in July; occasionally there is a small second emergence in late August. Rests concealed amongst reeds by day and flies at dusk and after dark. Males keep rather strictly to the close vicinity of reed-beds, often flying freely among reeds standing in open water. Females wander more widely. Both sexes come to light.

LARVA Feeds in stems of common reed (*Phragmites australis*) and reed sweet-grass (*Glyceria maxima*). On vacating a stem, it floats to another upon a raft made of a fragment of reed-stem. May and June. Pupa head upwards in a reed-stem, below an exit-hole made by the larva.

DISTRIBUTION Very local in large reed-beds, chiefly coastal, south of Yorkshire. Two records from Jersey (Long, pers. comm.).

Donacaula forficella (Thunberg)

Pl.2, figs 28($\male$), 29($\female$); *frontispiece*, fig. 3

IMAGO Wingspan 25–30mm. Sexually dimorphic, though females are hardly larger than males. *Male*. Forewing whitish, ochreous tinted, with two small black discal dots, brownish subcostal streak and dark, oblique apical streak; hindwing white. *Female*. Forewing very obliquely pointed, termen concave, light ochreous with blackish discal dot and traces of dark subcostal and apical streaks frequently present, sometimes strongly developed; hindwing white. In *D. mucronellus* (Pl.2, fig.31), costal stripe is sharply demarcated from dark subcostal streak and apical streak is wanting. A melanic form occurs in bogs in the New Forest, Hampshire.

Single-brooded; flies in June and July. Readily disturbed from rank vegetation by day; both sexes fly naturally at dusk and after dark, and come to light.

LARVA In rolled shoots of common reed (*Phragmites australis*), reed sweet-grass (*Glyceria maxima*) and sedges (*Carex* spp.). When feeding on plants growing in open water, larva floats to a fresh plant on a raft, in the manner of *S. gigantella*. May and June. Pupa in a tough cocoon in the larval working.

DISTRIBUTION In reedy ditches, marshes, fens and bogs. Locally fairly common to common in England from Yorkshire and Lancashire southwards, in Ireland and the Channel Islands.

Donacaula mucronellus (Denis & Schiffermüller)

Pl.2, figs 30($\male$),31($\female$)

IMAGO Wingspan male 22–26mm, female 29–35mm. Forewing of female with pointed apex and oblique termen, but external differences otherwise less marked than in other Schoenobiinae. Forewing of male greyish ochreous, that of female more ochreous, each with broad, pale costal streak sharply demarcated from dark subcostal streak which extends from base to apex. Hindwings white, fuscous-tinted in male.

Flies in June and July, occasionally in late August. Rests concealed by day and is rarely seen. At dusk, males fly weakly amongst marsh vegetation. Both sexes come to light.

LARVA Feeds in lower part of stems of common reed (*Phragmites australis*), reed sweet-grass (*Glyceria maxima*) and sedges (*Carex* spp.) from July (and ? September) to May. Pupa in base of stem below a small, round exit-hole 50–75mm above root.

DISTRIBUTION In ditches, marshes and fens; local in the southern half of England, Perthshire and Aberdeenshire (Pelham-Clinton, pers. comm.), western Scotland including the Inner Hebrides, and Ireland. Also recorded from Bangor, Caernarvonshire (Agassiz, 1984).

Subfamily SCOPARIINAE (14 species)

The Scopariinae are a group of small, triangular-winged greyish moths which rest by day on tree-trunks or walls and rocks. Some are very easily disturbed and dash away when approached; others are more docile. The species are hard to differentiate, some exceedingly so, so care is needed when recording them. The larvae feed in the roots of herbaceous plants or on mosses.

For the most recent taxonomic work on this subfamily, see Leraut (1983).

Scoparia subfusca Haworth
= *cembrella* auctt.

Pl.3, figs 1–3

IMAGO Wingspan 20–27mm. Forewing whitish grey, rather evenly irrorate light brownish grey; first line curved, or more often obtusely angled at fold, whitish, shaded fuscous along distal margin; second line weakly sinuate, angled above middle, whitish, dark-edged basally; subterminal line invisible, or indicated by faint, pale, radiating interneural streaks; discal spot rather weak, X-shaped, the arms fading, but sometimes tending towards an 8-shape; orbicular stigma indistinctly outlined, oval, touching first line although sometimes only the distal part of the rim can be seen; claviform stigma short, blackish, touching first line or just extending basally through it; a series of weak, brownish terminal dots on ends of veins; fringe whitish, chequered brownish fuscous in basal half. Hindwing whitish grey, fringe whitish with very weak brownish subbasal line. In f. *scotica* White (fig.2), wings darker brownish grey, almost devoid of markings; specimens from Orkney and Shetland are small, dark, but with pale cross-lines fairly distinct (fig.3).

One of the larger Scopariinae, and usually easy to recognise by the rather uniform light brownish grey colour, simple markings and lack of distinct subterminal line. The contrast between the pale cross-lines and their darker edging is rather variable but never very striking.

Single-brooded; flies in June and July. Rests by day on rocks, stone walls and tree-trunks and flies off when disturbed, but seems to be encountered at rest less often than some of the other species. At night, it comes to light.

LARVA Feeds from September to May on roots of ox-tongue (*Picris* spp.) and sometimes colt's-foot (*Tussilago farfara*), channelling the surface of the root beneath a slight web.

DISTRIBUTION Fairly common to common locally throughout the British Isles; smaller and darker specimens tend to be found in north and west, and in mountains.

Scoparia pyralella (Denis & Schiffermüller)
= *arundinata* (Thunberg)
= *dubitalis* (Hübner)

Pl.3, figs 5–7

IMAGO Wingspan 17–20mm. Forewing whitish, rather heavily mottled or marbled with rich brown irroration, upon which the white cross-lines are conspicuous: the first rather oblique, curved, subsinuate, the second slightly angulate-sinuate above middle; discal spot 8-shaped, light brown, incompletely ringed dark brown, orbicular and claviform stigmata of similar colour, incompletely outlined dark brown, both touching first line; area between second line and termen brownish fuscous, traversed by rather narrow, irregular white subterminal line which reaches neither costa nor tornus, and very narrow, white, crenate terminal line extending from round apex to tornus; fringe fuscous at base, then whitish, then light fuscous brown. Hindwing whitish, rather glossy, darker towards margin; fringe white with darker basal line. In f. *ingratella* Zeller, Knaggs *nec* Zeller (fig.6), the dark brown marbling is very much reduced, leaving the darkly outlined stigmata and a few dark clouds along costa; in f. *purbeckensis* Bankes (fig.7), the dark markings are reduced to two fascia, one distal to the first line and embracing the claviform stigma, the other passing between the discal spot and second line; f. *alba* Tutt is almost pure white, while retaining the general 'jizz' of the species.

Although this is a variable species, it is usually easy to recognise by the rather short, broad wings, contrasting pattern and the distinctly ochreous rather than grey hue.

Flies in June. Rests by day chiefly on the ground or on stones, but also on rocks and tree-trunks; very easily disturbed, when it flies off wildly and in quite a different manner from its gentle excursions at dusk.

LARVA Feeds in April and May on decaying plant material in a slight web; probably also on roots of common ragwort (*Senecio jacobaea*) and other plants.

DISTRIBUTION Common throughout much of England, Wales and Ireland, more local in Scotland, preferring open country on light soils; most abundant on chalk downs, waste ground near the sea and on coastal shingle. f. *purbeckensis* is evidently confined to the Swanage district of Dorset, and f. *alba* to Kent, Surrey and Sussex. Recorded from the Channel Islands.

Scoparia ambigualis (Treitschke)

Pl.3, figs 9–11; *frontispiece*, fig.4; text figs.5a-c,l

IMAGO Wingspan 15–22mm. Forewing greyish white, often with faint ochreous hue, sprinkled dark fuscous especially along costa; first line oblique from costa, angled above middle, then usually doubly crenate to dorsum which it meets at right angles, whitish, dark-edged distally, dark edging thicker towards costa; second line sinuate, strongly angled above middle, whitish with darker clouds both basally and distally; subterminal line a wide, irregular whitish cloud which extends from near apex to tornus, touching second line at its angle; discal spot light ochreous, 8-shaped, blackish edging forming a distinct X; orbicular stigma oval, ochreous, dark-rimmed, touching first line; claviform stigma dark, usually with a small pale area between it and first line; terminal line pale, broken, contiguous with subterminal near apex and tornus; fringe whitish, incompletely divided by two fuscous lines.

47

Hindwing light greyish fuscous, darker towards margin; fringe whitish with broken subbasal line. *S. ambigualis* is very variable; the more extreme forms are, however, fairly easily recognised. The smaller, darker f. *atomalis* Stainton (fig.10) is characteristic of northern Britain; ab. *klinckowstromi* Hamfelt has dark basal and terminal areas of forewing contrasting with pale median area, while in ab. *crossi* Bankes the median area is dark and the basal and terminal areas pale; both are rare. In the whitish ab. *octavianella* Mann the dark markings are reduced, while in a curious form from boggy moorland at Malham Tarn, Yorkshire (fig.11), the ground colour is dark and the markings virtually absent. *S. ambigualis* is easily confused with *Eudonia truncicolella* (Stainton) (Pl.3, fig.16), but that species has more pointed forewing, the markings are blacker and the wings lack the faint ochreous wash found in *S. ambigualis*; the genitalia (text fig.5) show the differences of the genera: in *Scoparia*, male genitalia show long pointed uncus, prominent pointed process on ventral margin of valve, and cornuti in aedeagus, whereas in *Eudonia*, uncus is blunt, valve lacks spiny process (text fig.5j) and aedeagus is without cornuti (text fig.5k). Differences between *S. ambigualis* and *S. ancipitella* (de la Harpe) and *S. basistrigalis* Knaggs are given under those species, *q.v.*

S. ambigualis is the first scopariine to appear on the wing, in May and June, though f. *atomalis* does not fly until July or August. The typical form rests by day on tree-trunks, preferring oak, and f. *atomalis* on heather, stones and rocks. Both forms dash wildly from their resting place when disturbed by day but fly more gently at night, when they come to light.

LARVA The early stages appear to be unknown, though the larva is believed to feed on moss.

DISTRIBUTION Common throughout Great Britain and Ireland, including the Shetlands and Hebrides, in deciduous woodland in the south and on moors in the north and west. The Channel Islands.

Scoparia basistrigalis Knaggs

Pl.3, fig.4; text figs 5d–f,m

IMAGO Wingspan 20–23mm. This species is usually easy to recognise though some individuals are very close indeed to *S. ambigualis* (Pl.3, figs 9–11). In general, *S. basistrigalis* is a little larger, broader-winged, with apex of forewing blunter; greyer, without ochreous hue, more heavily irrorate blackish and with fringe more strongly chequered. In cases of doubt, genitalia should be checked: in male of *S. basistrigalis*, uncus is tapered to a narrow point, not exceeded by gnathos (text fig.5d), and aedeagus has two rows of about four cornuti (text figs e,f), whereas in *S. ambigualis* uncus is isosceles triangular and gnathos is longer (text fig.5a), and aedeagus has two or three strong, thorn-like cornuti, with three or more smaller ones at base (text figs 5b,c); in female of *S. basistrigalis*, bursa is covered with short spines on one side, squamous on other, and ductus, narrow at junction with ostium, widens posteriorly forming a bulb (text fig.5m), whereas in *S. ambigualis*, bursa is squamous and ductus is not bulbed (text fig.5l).

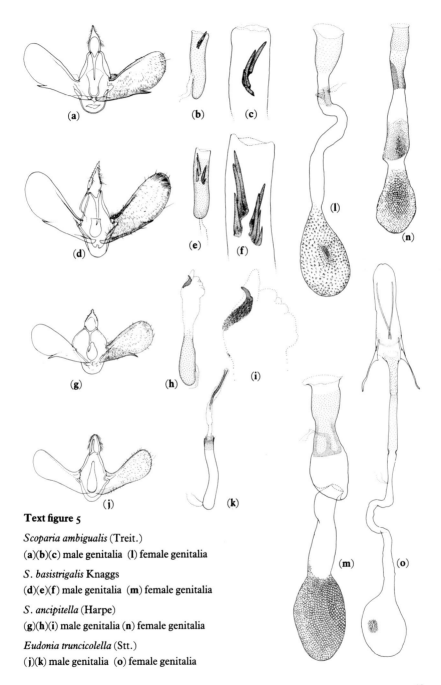

Text figure 5

Scoparia ambigualis (Treit.)
(**a**)(**b**)(**c**) male genitalia (**l**) female genitalia

S. basistrigalis Knaggs
(**d**)(**e**)(**f**) male genitalia (**m**) female genitalia

S. ancipitella (Harpe)
(**g**)(**h**)(**i**) male genitalia (**n**) female genitalia

Eudonia truncicolella (Stt.)
(**j**)(**k**) male genitalia (**o**) female genitalia

Single-brooded; flies in July. Rests by day on tree-trunks, head upwards, and is less ready to fly than others of the genus. It comes freely to light.

LARVA Biology apparently unknown.

DISTRIBUTION Occurs in deciduous woodland in southern England and is common, for instance, in the New Forest, Hampshire, the Chiltern beechwoods and at Chippenham Fen, Cambridgeshire. More local in northern England, and evidently unrecorded from other parts of the British Isles. Isles of Scilly, uncommon (Agassiz, 1981b).

Scoparia ancipitella (de la Harpe)
= *ulmella* Knaggs

Pl.3, fig.8; text figs 5g–i,n

IMAGO Wingspan 18–21mm. Superficially very similar to *S. ambigualis* (Pl.3, figs 9–11), *q.v.*; well-marked individuals can be identified with some confidence by the lack of ochreous tint, the strongly developed, dark, evenly curved first line, the more sinuate second line, and subterminal line which is approximated to the second throughout its length and often difficult to distinguish from it, and the paler hindwing. *S. ancipitella* is not so broad-winged, nor so irrorate blackish, as *S. basistrigalis* (Pl.3, fig.4). The genitalia are distinct: in the male, the uncus is more abruptly tapered than in *S. ambigualis* but lacks the long point characteristic of *S. basistrigalis*, shorter than gnathos (text fig.5g), and in the aedeagus there is one very characteristic short, thick, broad-based cornutus (text figs 5h,i); in the female, bursa has a large patch of spines anteriorly and the ductus is bulbed close to junction with ostium bursae, the bulb covered with small, rasp-like teeth (text fig.5n).

Single-brooded; flies in July and August. Rests by day on lichen-covered trunks of oak and elm and is easily alarmed; when flushed, it looks paler than other species on the wing. Flies at night and comes to light.

LARVA Biology apparently unknown.

DISTRIBUTION Imperfectly known; very rare in southern England and, though recorded first from Hampshire, it appears that this species has a broadly northern and western distribution from Somerset to Argyll but is less common in the Midlands; found not uncommonly in mid-Wales (pers. obs.).

Dipleurina lacustrata (Panzer)
= *crataegella* auctt.
= *centurionalis* sensu Beirne

Pl.3, fig.14

IMAGO Wingspan 16–18mm. Forewing whitish, usually with faint ochreous tint, sprinkled dark fuscous or black, basal area clouded with blackish irroration; first line irregular, whitish, black edged distally, second weakly curved above middle, feebly denticulate, clouded blackish on each side; broad white subterminal line is interrupted in middle and diverted into tornus by blackish clouding; a very fine pale line along termen, bordered distally by a row of blackish interneural dots; discal spot X-shaped; orbicular and claviform stigmata blackish; fringe whitish with broken, blackish subbasal line. Hindwing light fuscous, darker towards termen; fringe whitish grey with dark subbasal line. Similar to *Eudonia mercurella* (Linnaeus) (Pl.3, figs 20,21), but that species has slightly narrower, more pointed forewing, is usually darker coloured, and subterminal line is not interrupted.

Single-brooded; flies in June and July. Rests by day on tree-trunks and old stone walls in rather open country.

LARVA Feeds in March and April in a slight tubular web amongst mosses on old stone walls and tree-trunks.

DISTRIBUTION Widely distributed in England, Wales and south Scotland, in Ireland and in the Channel Islands.

Eudonia pallida (Curtis)

Pl.3, fig.12

IMAGO Wingspan 17–19mm. Forewing pale whitish ochreous, weakly sprinkled reddish dark brown; first line obsolete, second sometimes well-defined, white, weakly sinuate, sometimes with darker clouding distally; subterminal line obscure, wide, whitish, bordered along termen by a series of dark brown interneural dots; discal spot dot-like, star-shaped or elongate; orbicular and claviform stigmata variably expressed, often elongate; all stigmata dark brown or blackish. Hindwing whitish, greyer in female, with paler fringe. One of the smallest British scopariines, easily recognised by the weak markings, pale coloration, and small, blackish stigmata.

Single-brooded; flies in June and July; rests amongst vegetation by day and drops when disturbed. Flies just before dusk and after dark, and comes to light.

LARVA Unknown; supposed to feed on mosses or lichens growing on the ground in marshy or boggy situations.

DISTRIBUTION Occurs in marshes, fens and bogs, sometimes commonly. Widespread in Great Britain and Ireland. Two records from Jersey (Long, pers. comm.).

Eudonia alpina (Curtis)
= *borealis* (Tengström *nec* Lefebvre)

Pl.3, fig.13

IMAGO Wingspan 20–25mm. Forewing long and narrow, with arched costa and very oblique termen; grey, heavily irrorate light brownish ochreous or brownish and sprinkled black; first line obscurely greyish white, strongly sinuate, oblique, edged distally with blackish irroration; second line sometimes clear, whitish, variably sinuate, strongly notched near costa, edged on both sides with blackish dots and irroration; discal spot black, X-shaped, upper half often completely ringed; orbicular stigma ringed with black; claviform sometimes conspicuously black, or absent; fringe brownish grey with brown clouding. Hindwing greyish fuscous, sometimes with weak postmedian line and small discal dot; fringe slightly paler. Much the largest of the narrow-winged scopariines, with characteristically oblique termen and brownish hue. Variable in size and development of markings.

Single-brooded; flies in June and July, resting by day on mossy turf of high mountains; easily flushed. Also flies naturally in afternoon sunshine.

LARVA Early stages unknown.

DISTRIBUTION A species of high mountains in Scotland above c.700m and at lower altitudes in Orkney and Shetland, though, according to Lorimer (1983), confirmation of its occurrence in Orkney is needed. In Scotland it is known to be local and irregular in its appearance, sometimes occurring commonly then not encountered again for years.

Eudonia murana (Curtis)

Pl.3, fig.15

IMAGO Wingspan 18–22mm. Forewing rather pointed, costa weakly arched, chalk-white with trace of ochreous on costa, rather evenly sprinkled with blackish scales which impart a mottled appearance and obscure the sharp outlines of the markings; first line oblique, weakly zigzag though variable in shape, white, edged black especially along outer edge; postmedian line oblique, curved round discal spot and angled on to costa; subterminal line obscure, bent basad round dark, broadly triangular median terminal blotch; discal spot concolorous, 8-shaped, blackish edging forming a well-defined 8 or X; orbicular stigma oval, blackish edged or entirely dark; claviform stigma short, black; fringe shining white with fuscous chequering which forms a broken line. Hindwing light grey; fringe white with irregular, broken, fuscous subbasal line. *E. murana* and *E. truncicolella* (Stainton) (Pl.3, fig.16) are sibling species which are difficult to separate, though differences between series examined side by side are usually clear enough: *E. murana* is evidently less variable; the forewing is more mottled – in *E. truncicolella* the markings are 'scratchy' in comparison; the postmedian line (subterminal line *sensu* Whalley & Tweedie, 1963) is nearly straight from post-discal curve to dorsum in *E. murana* whereas in *E. truncicolella* it is zigzag. The genitalia are very similar (see Whalley, *loc.cit.*).

Emmet (1979) states that the species is double-brooded and gives June–July and August as flight times. Rests by day on rocks (cf. *E. truncicolella*). There is sometimes a strong pre-dusk flight. After dark the species comes to light.

LARVA Feeds from February to May and in July (Emmet, *loc.cit.*, quoting Beirne, 1952), on mosses growing on rocks and old walls. Pupa in a cavity under moss.

DISTRIBUTION Found locally on moors and mountains from Devon and Dorset through Wales, northern England and Scotland, up to at least 1000m on Cairn Gorm. Recorded from Hoy, Orkney (Lorimer, 1983).

Eudonia truncicolella (Stainton)

Pl.3, fig.16; text figs 5j,k,o

IMAGO Wingspan 18–23mm. Very similar to *E. murana* (Pl.3, fig.15) in general appearance; the irroration of forewing is more 'scratchy', less mottled in *E. truncicolella* and the markings are usually more distinct, showing greater contrast with the ground-colour; in all the better-marked individuals, the zigzag postmedian line appears to be the most reliable character for distinguishing *E. truncicolella* from *E. murana*. Differences from *Scoparia ambigualis* (Pl.3, figs 9–11) are given under that species, *q.v.*

Single-brooded; flies in July and August. Rests by day on tree-trunks (cf. *E. murana*), chiefly oak and pine, and is easily disturbed. Flies from dusk onwards into the night and comes to flowers and light.

LARVA Feeds from September to June in galleries amongst mosses growing on soil or on stones, living in a silken tube. Pupa in June and July, in moss.

DISTRIBUTION Locally common in large woods in most of mainland Britain and Ireland. Inner Hebrides (Wormell, 1983). Two records from Orkney (Lorimer, 1983).

Eudonia lineola (Curtis)

Pl.3, fig.17

IMAGO Wingspan 18–20mm. Forewing pointed, costa nearly straight, arched only at base and near apex; chalk-white, irregularly sprinkled with black and olive fuscous scales, the latter imparting a characteristic tint found in no other species within the subfamily; first line oblique, strongly notched at dorsum, white, distally edged black; second line sinuate, strongly indented basad as it curves round discal spot, edged on each side by patches of olive fuscous and blackish scales; discal mark black, X-shaped, the stigma itself inconspicuous; orbicular stigma clearly outlined black; claviform stigma short, solid, black; terminal area largely white, marked by a series of black dots and, in middle region, short dashes; fringe white with broken dark lines at tip and near base. Hindwing whitish, irrorate grey, with moderately conspicuous dark discal spot; fringe white with traces of dark terminal and subbasal lines. *E. lineola* is a narrow-winged species distinguished by the presence of clear markings on olive-tinted ground-colour, the shapes of the cross-lines and by the presence of a discal spot in the hindwing.

Single-brooded; flies in July and August. Rests by day on lichen-covered branches or fences and is easily disturbed. Flies at dusk and after dark and comes to light.

LARVA Feeds in spring, probably from previous autumn, on lichens (*Parmelia* spp.) growing on fences and old branches, especially those of hawthorn and blackthorn, and on old walls and rocks. Pupa in June and July in a slight cocoon under lichen.

DISTRIBUTION A very local and usually uncommon species but widely distributed in Britain and Ireland, chiefly in the coastal counties from Aberdeenshire to the Isle of Wight, and from the Burren, Co. Clare. Very common in Guernsey, but only one record from Jersey (Long, pers. comm.).

Eudonia angustea (Curtis)

Pl.3, fig.18

IMAGO Wingspan 17–22mm. Forewing pointed, costa nearly straight, termen very oblique, chalk-white sprinkled with fuscous and brown scales; base of wing brownish; first line broad, oblique, coarsely serrate, white, edged distally with blackish; second line strongly sinuate, strongly indented basad beyond discal mark; discal mark X-shaped, the upper half filled with light brown; orbicular stigma narrowly oval, light brown, irregularly edged black; claviform stigma short, black; subterminal area white with row of black terminal spots; fringe white with narrow brown line. Hindwing pale whitish grey; fringe white with faint darker line. Easily recognised by the narrow, pointed, brown-tinted forewing.

Flies from July to October, sometimes much earlier in coastal localities. Rests by day on walls and fences and is more sluggish than most of the other Scopariinae. Flies at dusk and after dark, and comes to light and flowers.

LARVA Feeds in cushions of moss (*Tortula ruraliformis*) growing on sandhills and in moss on walls, making silken galleries down in the tufts. Found in spring and early summer. Pupates in a slight cocoon amongst moss.

DISTRIBUTION Local but widespread through the whole of the British Isles including the Shetlands and the Channel Islands; chiefly coastal.

Eudonia delunella (Stainton)
= *vandaliella* (Herrich-Schäffer)
= *resinella* auctt.

Pl.3, fig.19

IMAGO Wingspan 17–18mm. Forewing white, shaded light ochreous on costa, in disc and on dorsum; first line irregularly curved, white, edged on each side by blackish irroration; second line sinuate, white, contiguous with white subterminal line in middle, and separated from it at costa and dorsum by irregularly triangular blackish patches; discal spot large, quadrate, black, the area between it and costa filled with dark fuscous; orbicular and claviform stigmata black, included within irroration distal to first line; terminal spots large, blackish, more or less united to form an irregular line; fringe whitish, ochreous-tinged, with broken fuscous subbasal line. Hindwing light grey with faint rounded discal spot, darker towards termen; fringe whitish with dark subbasal line. An easily-recognised species with strongly contrasted black-and-white pattern and prominent discal blotch.

Single-brooded; flies in July and August. Rests by day on tree-trunks, especially those of ash. Flies from dusk into the night and comes to sugar and light.

LARVA Feeds in spring on lichens and mosses on trunks of ash, apple and sometimes elm.

DISTRIBUTION Local but widely distributed in woods in England, Wales, southern Scotland and Ireland, though the only recent records known are from the New

Forest, Hampshire, and Faringdon, Berkshire (Agassiz, 1983), and from Carmarthenshire, Dumfriesshire and south Devon (Pelham-Clinton, pers. comm.).

Eudonia mercurella (Linnaeus)

Pl.3, figs 20,21

IMAGO Wingspan 16–19mm. Forewing greyish white, more or less heavily suffused fuscous and black; first line curved, whitish irrorate fuscous, black-edged distally; second line fine curved outwards from costa and dorsum to a point above the middle where it touches subterminal line of similar colour and thickness which curves basad from apex and tornus – in darker specimens these lines form a distinctive wavy X-shaped configuration which occupies the whole of the subterminal region; X-shaped discal mark and dark orbicular and claviform stigmata are usually concealed in dark irroration in median area of wing; termen narrowly black-edged at base of fringe, which is white with broken fuscous subbasal line. Hindwing grey, fringe light grey with continuous fuscous subbasal line. A variable species, but easily recognised in all its forms by the dark coloration and characteristically shaped cross-lines; hindwing darker than that of any other British scopariine species, except the much larger and differently patterned *E. alpina* (Pl.3, fig.13). The f.loc. *portlandica* Curtis (fig.21) from the Isle of Portland, Dorset, has whitish base to forewing.

Single-brooded: flies from June to September. Rests by day on rocks, walls and tree-trunks and is easily disturbed. At night, it comes freely to light.

LARVA Feeds in spring in tufts of moss growing on tree-trunks, roots, rocks and walls, living in a silken gallery. Pupa in May and June, amongst moss.

DISTRIBUTION Common and widespread in England, Wales, southern Scotland, Hebrides, Ireland, and the Channel Islands.

Subfamily NYMPHULINAE (15 species)

The Nymphulinae are the familiar, long-legged and distinctively patterned chinamark moths which can be disturbed by day amongst waterside vegetation. Their larvae are aquatic, living submerged in spinnings amongst water-weeds. Several tropical species have appeared as adventives in greenhouses specializing in aquarium plants.

Elophila nymphaeata (Linnaeus)　　　　　　Brown China-mark

Pl.3, figs 23,25(♂♂), 24(♀); *frontispiece*, fig.5

IMAGO Wingspan 25–33mm. Forewing whitish suffused various shades of light brown or fuscous; basal area with obscure dentate white and dark fuscous cross-lines; first and second lines obscure, white, dark-edged, the first angulated above

55

middle, second abruptly sinuate inwards below middle; three large, white, brown-edged blotches in median area; white subterminal band variably interrupted and cut by dark veins. Hindwing white with broad, irregular antemedian and postmedian brown bands and large, kidney-shaped discal spot which is often united to postmedian band; a narrow, irregular, white subterminal band and narrow, golden, brown margined terminal line. Differs from *Nymphula stagnata* (Donovan) (Pl.3, figs 32,33) in larger size and broader, browner wings. A particularly dark brown form occurs in the bogs of the New Forest, Hampshire (fig.25).

Single-brooded; flies from late June to August. Readily disturbed by day from amongst waterside vegetation; flies freely at dusk and later comes to light.

LARVA Aquatic; feeds on pondweeds (*Potomogeton* spp.), frogbit (*Hydrocharis morsus-ranae*), bur-reeds (*Sparganium* spp.) from September to June, at first mining a leaf, later living in a floating case constructed of fragments of leaf; when feeding, it attaches the case to the underside of a leaf with silk. Pupates in a silk cocoon covered with pieces of leaf, attached to the stem of a water-plant just below the surface.

DISTRIBUTION Around margins of ponds and lakes, and along stretches of slow flowing rivers and canals amongst lush vegetation. Common in suitable localities through most of the British Isles including the Channel Islands.

Elophila difflualis (Snellen)
= *enixalis* (Swinhoe)

Pl.3, figs 26(♂), 27(♀)

IMAGO Wingspan of male 11–13mm, female 13–18mm. Sexes similar in pattern though differing considerably in size, and female is less brightly marked, with paler hindwing. Detailed descriptions are given by Agassiz (1978). This is one of several small nymphulines recently found infesting aquatic greenhouses in southern England, having been introduced with stock from the Far East or the U.S.A. *E. difflualis* can be distinguished from *E. melagynalis* (Agassiz) (Pl.3, figs 28,29) by its slightly larger size, paler coloration of the female, whitish subterminal line, by the presence of a diffuse pale patch in the discal area of the forewing and by the absence of orange-yellow discal spots on upper- and underside of fore- and hindwing, present in *E. melagynalis*. *E. manilensis* (Hampson) (Pl.3, figs 30,31) is a more robust, heavier-bodied insect, darker in colour with forewing more uniform fuscous brown and with diffuse ferruginous discal spot. *Synclita obliteralis* (Walker) (Pl.4, figs 10,11) is a darker species, and hindwing especially is much darker. In *Parapoynx obscuralis* (Grote) (Pl.4, fig.1), *P. diminutalis* (Snellen) (Pl.4, figs 2,3) and *P. crisonalis* (Walker) (Pl.4, fig.6), the wings are more elongate, there is a dark discal spot in forewing and markings in hindwing are much more oblique.

LARVA When young, inhabits a tight spinning among leaves of pondweeds (e.g., *Vallisneria, Synnema, Echinodorus, Potomogeton, Marsilia*), later living in a floating case made of leaf-fragments.

DISTRIBUTION *E. difflualis* was first discovered in Britain in 1977, in aquatic nurseries at Enfield, Middlesex, which have since proved a rich hunting-ground for these tropical nymphulines. *E. difflualis* is a native of the Far East, where it is widespread. In the greenhouses, the moth rests head downwards with abdomen raised and wings arranged 'tent fashion' over the body. At Enfield, it has been by far the commonest species recorded.

Elophila melagynalis (Agassiz)

Pl.3, figs 28(♂), 29(♀)

IMAGO Wingspan of male 9–10mm, female 10–12mm. The smallest nymphuline found in aquatic nurseries in southern England. The species is described by Agassiz (1978). Most like a small *E. difflualis*, and differences are given under that species. A characteristic feature of the female is an almost vertical white bar on the spread hindwing.

LARVA Early stages unknown.

DISTRIBUTION Recorded uncommonly in the greenhouses at Enfield. Evidently a native of Sri Lanka, Indonesia and other parts of the Far East. Like *E. difflualis*, rests head downwards with abdomen raised, looking shorter-winged than that species.

Elophila manilensis (Hampson)

Pl.3, figs 30(♂), 31(♀)

IMAGO Wingspan of male 11–14mm, female 16–22mm. Full descriptions of both sexes are given by Agassiz (1981a). Forewing of both sexes rather uniformly dark fuscous brown with diffuse ferruginous discal spot. Hindwing of female strongly patterned, but whiter than that of male. The darkest and most heavily built of the nymphulines found in aquatic greenhouses.

LARVA Polyphagous on aquatic plants, often living fully submerged; when young it lives in a tight spinning, but later in a floating case made from leaf-fragments.

DISTRIBUTION Discovered in 1978 in the greenhouses at Enfield. A native of the Far East, recorded from India, Burma, Thailand, the Philippines, and throughout Malaysia and Borneo, though not, apparently, from Sri Lanka. Male rests in the same fashion as the other *Elophila* species, but female is more secretive.

Synclita obliteralis (Walker)

Pl.4, figs 10(♂),11(♀)

IMAGO Wingspan of male 11–13mm, female 13–17mm. Female is larger and less brightly marked than male. A detailed description of both sexes is given by Shaffer (1968b). In both sexes, the basal half of the forewing is distinctly darker than the proximal half, and the hindwing is more uniformly coloured than in any other of the adventive nymphulines. The male is blacker, less brown, than any of the others, female is lighter brown, of similar shade to *Elophila manilensis* (Pl.3, fig.31).

LARVA Aquatic, gill-less; on various aquatic plants; in North America commonly on water-lilies, sometimes as a greenhouse pest, and also abundantly on duckweed (*Lemna* spp.) (Munroe, pers. comm.).

DISTRIBUTION Introduced with water plants from the U.S.A. to aquatic green-houses at Enfield, and the first to be recorded there. Endemic to the United States, from Texas to Florida and the eastern states, and introduced to the Hawaiian Islands. There is some evidence that it can spread from greenhouse to greenhouse in Britain, but none so far that it can survive in the wild.

Nymphula stagnata (Donovan)   Beautiful China-mark
Pl.3, figs 32,33

IMAGO Wingspan 20–25mm. Sexes similar, though female often a little larger. Forewing shining white, with delicately pencilled brown cross-lines and outlines of large rounded markings in median area of wing. Hindwing white with brown antemedian and postmedian lines and linear crescentic discal spot which touches antemedian line. Sometimes markings on all wings are very faint (fig.33). A smaller and whiter species than *Elophila nymphaeata* (Pl.3, figs 23–25); lightly marked specimens could be mistaken for male *Parapoynx stratiotata* (Linnaeus) (Pl.3, fig.34), but in that species the forewing discal spot is a closed circle and hindwing lacks antemedian line, postmedian line is thicker and broken, and discal spot is a dark point.

Single-brooded; flies in July and August. Easily disturbed by day from amongst lush waterside vegetation; flies in evening low over surface of water; comes to light and may travel some distance from its breeding ground.

LARVA Aquatic. Feeds from August to May, chiefly on bur-reeds (*Sparganium* spp.) but also on yellow water-lily (*Nuphar lutea*) and probably other water-plants. When young it enters the stem of the foodplant, mining the pith in all directions. Hibernates in the stem and recommences activity in April, feeding either in the stem or in a chamber of spun leaves just below the surface of the water. Pupates in a white silken cocoon attached to the foodplant.

DISTRIBUTION Margins of rivers, streams and lakes and common in such habitat throughout England, Wales, Ireland and southern Scotland, more local in central and north Scotland, Inner Hebrides (Wormell, 1983). The Channel Islands.

Parapoynx stratiotata (Linnaeus)   Ringed China-mark
Pl.3, figs 34(♂),35♀)

IMAGO Wingspan of male 22–24mm, female 28–30mm. Sexually dimorphic, fore-wing of male shorter and more rounded than that of female, and hardly exceeding length of hindwing. Male forewing whitish with ochreous gloss, sparsely irrorate light brown; first line white, irregularly brown-edged distally, curving strongly basad to costa; second line pure white, broadly brown-edged basally, sinuate about middle; discal spot prominent, white, brown-edged; fringe weakly chequered. Hindwing white with broken, brown postmedian line and small dark discal spot. Forewing of female heavily suffused fuscous especially along veins; cross-lines obscure, discal spot remaining prominent, white, blackish-edged. Distinguished from other species by the white, dark-rimmed discal spot, present in both sexes.

Single-brooded; flies from June to August. Male is easily disturbed by day from waterside vegetation, but female is seldom seen until nightfall. Both sexes fly freely from dusk onwards and come to light; female especially is sometimes taken some distance from breeding grounds.

LARVA Structurally the most specialised of the British nymphulines for an aquatic way of life. Feeds on pondweeds (*Potomogeton* spp.), Canadian waterweed (*Elodea canadensis*), hornwort (*Ceratophyllum* spp.) and other water-plants from July to May, spinning the leaves together and living in an open web. Makes periodic wriggling movement of body as an aid to gas exchange. Pupates in a large, pinkish, subaquatic cocoon attached to a plant.

DISTRIBUTION Local in England from Yorkshire and Lancashire southwards, commoner in the south; widespread and locally abundant in Ireland. Found around margins of ponds and lakes and along slow-flowing rivers, canals and drainage-ditches.

Parapoynx obscuralis (Grote)

Pl.4, fig.1

IMAGO Wingspan of male 14–16mm, female 17–19mm. Forewing of female longer, relatively narrower and more pointed than that of male. The species is described by Shaffer (1968a). The hindwing in both sexes is predominantly white, and its most obvious feature is a broad, blackish postmedian band which is evenly curved and runs nearly parallel to termen; basal to this, two weak, narrow, grey lines cross the white background, each nearly straight, very oblique, and lying almost parallel to the postmedian line; there is no discal spot, but a small dark mark in base of wing. These features serve to differentiate the species from *P. diminutalis* (Snellen) (Pl.4, figs 2,3), in which the cross-lines on hindwings are sigmoid, and there is a distinct dark discal spot in the majority of individuals. In *P. fluctuosalis* (Zeller) (Pl.4, figs 4,5), outer part of forewing is banded with orange, and there is a clear white subterminal stripe; hindwing is crossed by three weakly sigmoid dark fasciae, strongest of which is the outermost: the basal area therefore does not appear so white as in *P. obscuralis*. The indentation below the apex of hindwing of *Oligostigma* species, *q.v.* is sufficient in itself to separate them; and *O. bilinealis* Snellen (Pl.4, fig.8) is, in any case, the only one so far reported in Britain that has bands on the hindwing which follow approximately the same course as those of the *Parapoynx* species under discussion.

LARVA Aquatic, gilled. Feeds on water-plants.

DISTRIBUTION *P. obscuralis* was the first of the aquatic greenhouse adventives to be reported in Britain, from a nursery in Hemel Hempstead, Hertfordshire, in 1967; subsequently found at Enfield in 1984; introduced with water-weeds from Maryland, U.S.A.

Parapoynx diminutalis (Snellen)

Pl.4, figs 2($\circlearrowright$),3($\female$)

IMAGO Wingspan of male, 13–16mm, female 17–19mm. Females longer-winged, paler and more weakly marked than males. A detailed description is given by Agassiz (1978). Somewhat variable; distinguished from *P. obscuralis* under that species, *q.v.*

LARVA Aquatic; feeds on various species of water-weed.

DISTRIBUTION Reported in abundance from aquatic nurseries at Enfield in 1977, and subsequently, though less commonly, in the same nurseries and elsewhere. Numbers are checked by periodic fumigation of the greenhouses. A native of the Far East and Africa, where it is widespread. Enfield stock probably originated from Singapore. Occurs as an adventive in many parts of Europe and the U.S.A.

Parapoynx fluctuosalis (Zeller)

Pl.4, figs 4($\circlearrowright$),5($\female$)

IMAGO Wingspan 15–24mm. Female considerably larger than male, forewing more pointed, but coloration similar. Described in full by Agassiz (1981a). The only species with a clear, narrow white band in termen of forewing; hindwing more heavily banded than in other species of *Parapoynx*.

LARVA Biology unknown.

DISTRIBUTION A few specimens taken in the aquatic greenhouses at Enfield, first on 11 July 1979. Widespread in the Far East where it may, in fact, prove to be a complex of species.

Parapoynx crisonalis (Walker)

= *stagnalis* sensu Agassiz

Pl.4, fig.6

IMAGO Wingspan 11–15mm. Sexual dimorphism not pronounced. A small, pale species with small fuscous spot near base of forewing and double discal spot; hindwing with discal spot and weak antemedian fascia, strongest towards dorsum. Described by Agassiz (1981a) as *P. stagnalis* (Zeller) and later corrected (Agassiz, 1982).

LARVA Biology unknown.

DISTRIBUTION One specimen taken in the Enfield nurseries, 25 May 1979. Widespread and common in the Indo-Malayan region.

Oligostigma polydectalis Walker

Pl.4, fig.9

IMAGO Wingspan 13–17mm. Described by Agassiz (1981a). Another unmistakable species. Forewing white, marked with black and yellow; hindwing white with black Y-shaped median fascia and yellow, dark-edged subterminal fascia.

LARVA Biology unknown.

DISTRIBUTION Four specimens recorded from the nurseries at Enfield, 26 October 1978, 15 March 1979, 25 May 1979 and 7 June 1979. Originally described from Australia, and known to range from there through Malaysia.

Oligostigma angulipennis Hampson

Pl.4, fig.7

IMAGO Wingspan 14–17mm. Sexes similar; female slightly larger and darker. Described in detail by Agassiz (1978). An unmistakable species.

LARVA Inhabits a wet spinning between two leaves of water-plants (*Cryptocoryne* spp.). Pupation occurs within the spinning. Evidently continuous-brooded.

DISTRIBUTION Moderately common in the aquatic greenhouses at Enfield; first recorded 15 July 1977. Another native of the Far East, recorded from India and Sri Lanka. The Enfield stock originated from the latter locality.

Oligostigma bilinealis Snellen

Pl.4, fig.8

IMAGO Wingspan 14–17mm. Described in detail by Agassiz (1978). Characterized by pale orange ground-colour of forewing and strong, very oblique, white, fuscous-edged stripes on fore- and hindwings.

LARVA Biology unknown.

DISTRIBUTION A single dead specimen found in a strip-light insect-trap at an aquatic nursery in Hampshire in 1977 by the late D. W. H. Ffennell. A second specimen was taken at Bolton, Lancashire, 26 August 1983 (Hancock, 1984; pers. comm.). A native of the Far East.

Cataclysta lemnata (Linnaeus)   Small China-mark

Pl.4, figs 12(♂),13(♀)

IMAGO Wingspan of male 18–19mm, female 22–24mm. Sexually dimorphic, female larger and longer-winged than male. *Male.* Forewing shining white with dark-edged costa, interrupted brown subterminal line close to brown terminal line; very faint traces of other cross-lines; discal spot a minute dark point; fringe chequered. Hindwing white with conspicuous black band containing four bluish silvery dots,

61

which lie in termen between closely approximated brown subterminal and terminal lines; a dark discal spot, and traces of other cross-lines; fringe white with a broken dark band. *Female*. Forewing suffused brown, cross-lines obscure, dark discal spot present. Hindwing similar to that of male but smokier, with broad, shadowy antemedian line which extends from dorsum two-thirds of distance to costa, and larger discal spot. The characteristic black, silver-spotted band on hindwing of both sexes distinguishes this from other nymphulines.

Single-brooded; flies from June to August. Male is easily disturbed by day from amongst waterside vegetation, and at dusk flies low over water. Female is sluggish by day, but at night flies higher than male and wanders from breeding-grounds, and comes to light.

LARVA Feeds on duckweed (*Lemna* spp.), hibernating when young in a case made of fragments of foodplant; it may feed from time to time, but growth is completed in spring inside a larger case containing air, and made from an accumulation of overlapping duckweed thalli. Larva protrudes its head from this case in order to feed, and only in first instar is body of larva in contact with water. Pupation occurs just below the surface in a dense white silken cocoon amongst the foodplant.

DISTRIBUTION Characteristic of rather open habitat, ponds, ditches and canals and open parts of reed-beds where there is a good growth of duckweed, and common in such places from south Scotland through England, Wales and Ireland; the Channel Islands.

Subfamily ACENTROPINAE (1 species)

Recent research (Speidel, 1984) has shown *Acentria* to be unrelated to the Schoeno-biinae but to belong near the Nymphulinae or perhaps within that subfamily. Until agreement has been reached, I have decided to follow Leraut (1980) in placing the genus in its own subfamily but have transferred it to the proximity of the Nymphuli-nae. The narrow-winged, greyish white males of this curious species sometimes swarm at light on hot summer nights; the females are dimorphic and one form is apterous and completely aquatic, copulating at the surface. The larva is sub-aquatic.

Acentria ephemerella (Denis & Schiffermüller)   Water Veneer
= *nivea* (Olivier)

Pl.3, fig.22(♂)

IMAGO Wingspan 13–17mm. Female usually with rudimentary wings, though a fully-winged form occurs which is larger than the male. Forewing light grey, costa and veins brownish. Hindwing white. Quite unlike any other species.

Flies in an extended emergence from July to September, though life of an individual is no more than two days. Males fly at night low over water, touching it and leaving a wake as does a whirligig beetle (*Gyrinus*); on certain hot nights in summer they swarm, and appear in numbers at light far from water (cf. Hemiptera: *Corixidae*). Females usually remain below the surface of water, swimming actively by means of long-fringed middle and hind legs, and pairing occurs on the surface of water. The fully winged form of the female has been taken at light.

LARVA Subaquatic at depths of up to 2m; inhabits loose spinnings amongst the foodplants, chiefly Canadian waterweed (*Elodea canadensis*), pondweeds (*Potomogeton* spp.) and also stoneworts (*Chara* spp.) and filamentous algae. Hibernates from October to May, becoming full-fed in May or June. Pupation occurs in a silken, subaquatic cocoon up to a metre below the surface.

DISTRIBUTION Locally abundant in ponds, lakes and slow rivers throughout Britain and Ireland.

Subfamily EVERGESTINAE (3 species)

The Evergestinae are distinguished from related subfamilies by the structure of the male genitalia. The three British species are medium-sized, creamy yellow insects which resemble Pyraustinae in their broadly triangular forewings and slender, brittle legs. The larvae are associated with Cruciferae.

Evergestis forficalis (Linnaeus)   Garden Pebble

Pl.4, fig.14

IMAGO Wingspan 27–31mm. Forewing whitish ochreous, weakly suffused rusty brown towards apex; two dark brown transversely placed discal spots, the dorsal the larger; numerous fine, oblique, weakly curved or undulate cross-lines of which that from apex to dorsum is the most conspicuous. Hindwing whitish with ochreous grey postmedian line which is often broken into a series of separate spots. An unmistakable species.

At least two broods, in May and June, and August and September, with sometimes a third generation in late autumn. Rests amongst herbage by day and does not fly readily, but becomes active at dusk and later comes to light.

LARVA Feeds in summer and autumn on various Cruciferae, especially cultivated varieties of *Brassica*, turnips and horseradish, and may be a pest of gardens and allotments. Pupa in a strong silken cocoon below ground; autumn larvae hibernate full-fed in their cocoons, pupating in spring.

DISTRIBUTION A common moth of gardens and allotments through the whole of mainland Britain and Ireland, and also in the Inner and Outer Hebrides, Isle of Man, the Isles of Scilly and the Channel Islands. Far less common away from human dwellings, even amongst cruciferous crops. Known to migrate, and has been taken at a lightship well out to sea.

Evergestis extimalis (Scopoli)

Pl.4, fig.16

IMAGO Wingspan 27–31mm. Forewing pale ochreous yellow, cross-lines faintly indicated by series of small, rusty dots; a short, oblique, rusty brown apical streak continuous with terminal suffusion of same colour; cilia fuscous. Hindwing nacreous white with brownish marks along termen and in fringe. Impossible to confuse with any other species.

Single-brooded; flies in June and July. Occasionally flushed by day from among foodplants, but more often encountered at dusk or at light.

LARVA Feeds on seed-heads of Cruciferae, especially perennial wall-rocket (*Diplotaxis tenuifolia*) along the Thames estuary, and charlock (*Sinapis arvensis*) and white mustard (*S. alba*) in Breckland, making a silken web in which several larvae may be found together. When full-fed, larva makes a tough cocoon in soil in which it hibernates, pupating in May.

DISTRIBUTION Of regular occurrence in Breckland and along the Thames estuary, otherwise an irregular immigrant and colonist in south-east England, most often in calcareous localities. Recorded twice from Jersey (Long, 1967).

Evergestis pallidata (Hufnagel)

Pl.4, fig.15; *frontispiece*, fig.6

IMAGO Wingspan 24–29mm. Forewing pale glossy straw-yellow; cross-lines, outlines of stigmata and veins, especially distally, rusty dark brown; a shade of similar coloration is variably developed between postmedian and subterminal lines. Hindwing nacreous white with incomplete greyish brown postmedian line and terminal line; fringe white marked brown. The reticulate effect of dark cross-lines and veins make this an easy species to recognise.

Single-brooded; flies from June to August. Easily disturbed in daytime, flies at dusk and comes to light.

LARVA Feeds in August and September on Cruciferae, chiefly winter-cress (*Barbarea vulgaris*). Gregarious, larvae resting side by side on a leaf when not feeding. Larvae hibernate when full-fed in subterranean cocoons, and pupate in spring.

DISTRIBUTION Locally fairly common in damp open woodland and marshy places in England as far north as Lincolnshire, in south-west Scotland and in Ireland. Recorded from Mull of Kintyre (Agassiz, 1984); also from Guernsey. In north America, where it is known as the purple-backed cabbageworm, it is a minor pest of a number of cultivated crucifers and also feeds on various wild species (Munroe, pers. comm.).

Subfamily ODONTIINAE (2 species)

The Odontiinae are poorly represented in Britain; indeed, there is but one species known to be resident. The subfamily is designated on genitalic characters, but the British native species is unlike any other pyrale in our fauna. Larvae are internal feeders in herbaceous stems and leaves, sometimes causing blister-mines.

Cynaeda dentalis (Denis & Schiffermüller)

Pl.4, fig.17; *frontispiece*, fig.7

IMAGO Wingspan 24–29mm. Forewing creamy white shaded with light olive-brown; postmedian line very strongly and irregularly dentate; veins distally white; fringe chequered: a fan-like pattern is produced by the combination of these markings, which is highly characteristic. Hindwing whitish, more or less suffused fuscous; regularly dentate postmedian line, with teeth extending towards termen along veins, shows clearly on paler specimens.

Single-brooded; flies in July. Rests among foodplant with wings steeply arched and flies reluctantly. At night it flies freely and comes to light, and may be found resting openly on the foodplant.

LARVA Feeds internally in lower part of stem and leaf bases of viper's-bugloss (*Echium vulgare*), causing the leaves to wither; occurs in May and June, probably following hibernation. Pupation occurs in a hard cocoon covered with fragments of dried leaf, on the surface of one of the dead leaves.

DISTRIBUTION Extremely local in south-east England, in coastal localities. Fairly common at Dungeness, Kent, and fairly regularly seen at Portland, Dorset; recorded also from Suffolk and other parts of the south coast west to Devon.

Metaxmeste phrygialis (Hübner)

Pl.4, fig.18

A single specimen supposed to have been taken in the Highlands of Scotland by the professional collector, Charles Turner, who died in 1868. It is a dark-coloured species which could be mistaken for *Psodos coracina* (Esper) (Geometridae). It is common at high altitudes in the Alps and also occurs in Scandinavia. Like *Psodos* species, it flies by day (Mason, 1892; Cockayne, 1953.)

Subfamily GLAPHYRIINAE (1 species)

Only one specimen of this widespread tropical subfamily has been reported in Britain. The Glaphyriinae have characteristic spatulate scales on the upper surface of the hindwings.

Hellula undalis (Fabricius)   Old World Webworm

Pl.4, fig.19

IMAGO Wingspan 17–19mm. Forewing rather pointed, light brown; subbasal line indistinct, pale, dentate; first line rather broad, whitish, irregularly dentate; narrowly edged on each side by darker suffusion; second line irregularly sinuate dentate, edged basally by a narrow brown line which widens at costa to form a distinctive brown spot; distal to costal extremity of this line is a larger, subapical, triangular brown blotch; termen with a series of blackish semilunar interneural dots; discal spot, large, reniform, oblique, sometimes encircled by a narrow white outline. Hindwing glossy white, variably suffused brownish, especially towards margin, veins darker. The oblique brown reniform mark and the brown markings on costa in region of second line are characteristic features.

One specimen taken in vicinity of a light-trap, East Prawle, Devon, 28 September 1967 by E. J. Hare (Shaffer, 1968a). Widespread throughout the tropics of the Old World and introduced accidentally into Hawaii (Munroe, pers. comm.).

Subfamily PYRAUSTINAE (53 species)

The Pyraustinae are mostly rather broad-winged, strongly patterned pyrales with long, brittle legs and a jerky flight. They are among the most familiar British species and represent about one-quarter of the pyralid fauna of the country. There are many tropical species, some of which are occasional migrants here. The larvae are mostly agile web-spinners or leaf-rollers on herbaceous plants.

Pyrausta aurata (Scopoli)

Pl.4, fig.20

IMAGO Wingspan 15–18mm. Forewing deep purple, suffused blackish; a round golden spot in postmedian region towards apex is often the only marking, though there is frequently a small gold patch between this and costa, sometimes a much smaller golden discal spot and weak indications of other spots forming postmedian line and basal patch. Hindwing blackish with curved golden yellow postmedian line, unbroken but often failing to reach margins; cilia whitish. Distinguished from *P. purpuralis* (Linnaeus) (Pl.4, fig.21) and *P. ostrinalis* (Hübner) (Pl.4, fig.22) by darker, duller forewing with single conspicuous gold spot, and hindwing unmarked in basal half.

Double-brooded; flies in May and June, and in July and August. Flies actively in sunshine and also at night, when it comes to light.

LARVA When young, feeds on under surface of a leaf, later in a web amongst flowers of Labiatae: mint (*Mentha* spp.), marjoram (*Origanum vulgare*), calamint (*Calamintha* spp.), clary (*Salvia* spp.); June and July, and in the autumn. The autumn larvae hibernate full-grown in cocoons in old flower-heads, pupating therein in April.

DISTRIBUTION Commonest on chalk downland but also in open country on other soils, amongst waterside vegetation where water mint occurs, and sometimes also associated with cultivated mint in gardens. Patchily distributed in England, Wales, southern Scotland to the Inner Hebrides, and the Channel Islands.

Pyrausta purpuralis (Linnaeus)

Pl.4, fig.21; text fig.6a

IMAGO Wingspan 15–22mm. Forewing purple, weakly suffused greyish brown; an oval, ochreous basal patch edged by a narrow, oblique, dusky line; two very small, round bright yellow spots in disc are often absent; broad, bright yellow, wavy postmedian line may be complete or broken into three irregular spots of which the post discal is the largest; a weak, yellow subterminal streak is sometimes present. Hindwing blackish, with curved yellow postmedian line; basal half usually with one or two yellow spots or short, broad streaks; a rather indistinct, purplish subterminal band; inner half of fringe dark fuscous, outer half greyish white. This species has long been confused with *P. ostrinalis* (Pl.4, fig.22) and differences in the biology of the two have yet to be worked out. Differences between the imagines of the two have been well summarised by Holst (1978) and are as follows: seen together, *P. ostrinalis* has forewing distinctly narrower, dull reddish violet rather than purple, and the markings straw yellow, less bright than in *P. purpuralis*; on the underside, the pale subterminal line in *P. purpuralis* is nearly straight and fades before the costa, whereas in *P. ostrinalis* it curves to meet the postmedian line at the costa (text fig.6b). Differences in the genitalia are slight, of a comparative nature, and are also summarised by Holst.

Double-brooded; flies in May and June and in July and August; second brood specimens are larger and more brightly coloured. Flies by day as well as at night; comes to light.

LARVA Feeds between spun leaves of corn mint (*Mentha arvensis*) and thyme (*Thymus* spp.) in June and in the autumn; autumn larvae probably hibernate when mature and pupate in spring, but this fact appears to require confirmation.

DISTRIBUTION Widely distributed in rather open country, damp or dry, but preferring grassland on chalk or limestone. Throughout mainland Britain and Ireland, and reported from the Inner Hebrides (Beirne, 1952).

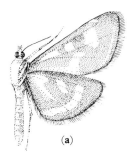

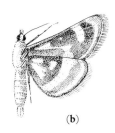

(a) (b)

Text figure 6 Underside (×3 approx.)

(a) *Pyrausta purpuralis* (Linn.) (**b**) *P. ostrinalis* (Hb.)

67

Pyrausta ostrinalis (Hübner)

Pl.4, fig.22; text fig.6b

IMAGO Wingspan 15–21mm. Forewing dull reddish violet, suffused blackish; markings similar to those of *P. purpuralis* (Pl.4, fig.21) but less bright, more straw yellow. Hindwing blackish with curved pale yellow postmedian line and usually a single large spot or streak in basal half towards costa; fringe similar to that of *P. purpuralis*.

The differences between *P. ostrinalis* and *P. purpuralis* are described under the latter species, *q.v.*

Double-brooded; flies at the same times and in the same localities as *P. purpuralis*.

LARVA and DISTRIBUTION The early stages have yet to be worked out, as have details of distribution. It appears to have a similar distribution to that of the preceding species; the two species have been seen flying together in localities as far apart as Bucks and the Burren, Co. Clare (pers. obs.). In Outer Hebrides (Waterston, 1981). Recorded from Guernsey. Very common in Jersey where apparently *P. purpuralis* is absent (Long, 1967).

Pyrausta sanguinalis (Linnaeus)

Pl.4, fig.23

IMAGO Wingspan 14–18mm. Forewing deep yellow; costa to two-thirds, oblique median fascia and termen crimson red; two small discal spots are just discernible within median fascia as it dilates towards costa. Hindwing grey, rather darker in female. Somewhat variable in the brightness and contrast of the markings, but unlikely to be confused with any other British *Pyrausta*.

Double-brooded; flies in June and in August in sunshine and also at night, when it comes to light.

LARVA Inhabits a silken tube amongst flowers of thyme (*Thymus drucei*) on which it feeds in July and in September and October. The autumn larvae hibernate in their tough cocoons and pupate in the spring. Cocoons are subterranean amongst moss and are usually attached to some solid object such as a stone.

DISTRIBUTION Very local but sometimes abundant where it occurs: on coastal sandhills in north Wales and Lancashire, in western Ireland, notably on the Burren limestone, and also recorded from Hartlepool, Cleveland, and Ayrshire.

Pyrausta cespitalis (Denis & Schiffermüller)

Pl.4, figs 24,25

IMAGO Wingspan 14–19mm. Forewing greyish or ochreous, heavily mottled darker grey or brown; small, dark reniform and orbicular stigmata and pale postmedian and subterminal lines usually obscure. Hindwing in male greyish ochreous, in female blackish, both with creamy postmedian and subterminal lines and basal half obscurely mottled, with dark discal spot placed towards costa. Forewing more mottled and less clearly marked than in any other British *Pyrausta* species; brightest forms occur in Shetland and the Isles of Scilly.

Double-brooded; flies in late May and June and again in July and August, in the sunshine and also at night, when it comes to light.

LARVA Gregarious in woven galleries at the base of leaves of plantains (*Plantago* spp.), in June and again in autumn. Pupa in a tough whitish cocoon in soil.

DISTRIBUTION On dry ground on sandy and calcareous soils, heaths, sandhills, cliffs and downs throughout Britain and Ireland, including the Shetlands, the Outer Hebrides and the Channel Islands.

Pyrausta nigrata (Scopoli)

Pl.4, fig.26

IMAGO Wingspan 14–17mm. Forewing black, weakly irrorate with rusty brown scales; markings creamy white: one or two minute round discal spots, narrow basal line from dorsum midway to costa, and rather thick postmedian line which bends in a double right angle towards dorsum. Hindwing black, with rather thick cream-coloured postmedian line which bends at an obtuse angle, and pale basal fleck. Cilia of both wings silky white. The only other black *Pyrausta* species in Britain is *P. cingulata* (Linnaeus) (Pl.4, fig.27) which is more sooty coloured, with narrower cross-lines which are nearly straight or curved, not angled. Features distinguishing these from the rare immigrant species, *Diasemia reticularis* (Linnaeus), *Diasemiopsis ramburialis* (Duponchel) and *Hymenia recurvalis* (Fabricius) (Pl.5, figs 31–33) are discussed under *Diasemia reticularis, q.v.*

Double-brooded; flies in June and July and in September and October. Diurnal.

LARVA Feeds on thyme (*Thymus drucei*), marjoram (*Origanum vulgare*) and also corn mint (*Mentha arvensis*) and woodruff (*Galium odoratum*), in a slight web under the leaves, in June and July and again in the autumn. Cocoon oval, whitish, surrounded by a layer of coarser silk.

DISTRIBUTION Local on rough downland slopes, especially where the ground is broken, in southern England and in Cumberland and Westmorland.

Pyrausta cingulata (Linnaeus)

Pl.4, fig.27

IMAGO Wingspan 14–17mm. Forewing sooty brownish black, unmarked save for narrow, weakly sinuate, creamy white postmedian line. Hindwing similarly coloured, with narrow, curved postmedian line. Cilia of both wings silky greyish white. Can be confused only with *P. nigrata* (Pl.4, fig.26), *q.v.*

Double-brooded; flies in May and June and again in July and August. Diurnal; flies along the edges of limestone cliffs and along the tops of thyme-covered walls.

LARVA Meadow clary (*Salvia pratensis*), given by most textbooks as the foodplant, evidently on the authority of C. von Heyden (Barrett, 1904, **9**: 176), is unlikely to be the normal foodplant in this country, where the moth appears to be associated with thyme (*Thymus drucei*). It is not known whether the larva has been reared in Britain. It should be sought within a silken web down amongst the foodplant near ground level, in June and in the autumn. Cocoon rather large, papery, greyish brown. Autumn larvae hibernate within the cocoon and pupate in spring.

DISTRIBUTION Local on chalk and limestone hills and cliffs, and on sandhills, walls and banks near the sea, in England, Wales and Scotland as far north as Perthshire and Argyll. In Ireland, common on the Burren, and recorded from Antrim, Louth and Sligo. Also from Guernsey and Jersey.

Margaritia sticticalis (Linnaeus)

Pl.4, fig.28

IMAGO Wingspan 24–29mm. Forewing fuscous; orbicular stigma oval, reniform stigma round, both a little darker than ground-colour and separated by a pale ochreous-white patch; postmedian line fine, whitish, broader at costa and angularly indented below middle; a more conspicuous pale terminal streak from apex to tornus; cilia a little darker than ground-colour, very glossy. Hindwing slightly paler than forewing, whitish towards costa, with darker postmedian line and indistinct pale terminal streak; fringe glossy grey, basal half dark fuscous. The larger size, pale terminal streaks and glossy cilia separate this species at once from forms of *Pyrausta cespitalis*; *Udea olivalis* (Denis & Schiffermüller) (Pl.5, fig.19) has a much whiter interstigmatal patch, lacks pale terminal streak and has whiter hindwing. No other predominantly fuscous-winged pyraustine has pale interstigmatal patch and terminal streak.

Flies in June, with a partial second generation in August and September. Readily disturbed by day in warm, sunny weather, but difficult for the eye to follow during its short flights close to the ground. Flies naturally at night and is a well-known migrant; comes to light.

LARVA Feeds on mugwort (*Artemisia vulgaris*), eating the upper sides of the leaves between the veins; June and July and again in September. Some pupae from the first generation emerge in late summer. Winter is spent as a pupa.

DISTRIBUTION Present status as a resident species uncertain, though for long a well-known inhabitant of the sandy Breck district of East Anglia. Otherwise a scarce migrant, chiefly in southern coastal counties from Suffolk to Cornwall.

Uresiphita polygonalis (Denis & Schiffermüller)
= *limbalis* (Denis & Schiffermüller)
= *gilvata* (Fabricius)

Pl.4, fig.29

IMAGO Wingspan 29–37mm. Forewing ochreous-brown, when fresh often mauve-tinted; small dot-like orbicular stigma and larger, obscurely outlined reniform stigma darker than ground-colour and usually clearly visible; other markings variably expressed: an oblique, weakly serrate antemedian line and postmedian line which is strongly curved from near middle to costa where it consists of a row of dots; the apical area and zone between the cross-lines may be suffused darker purplish brown. Hindwing orange-yellow with blackish border, a feature which renders the species unmistakable.

LARVA Apparently unrecorded in this country; on *Genista* and *Sarothamnus*.

DISTRIBUTION A very scarce immigrant in the southern seaboard counties in late summer and autumn. Taken at light. From central and south Europe through the tropics to Australia.

Sitochroa palealis (Denis & Schiffermüller)

Pl.4, fig.30

IMAGO Wingspan 29–34mm. Forewing pale sulphur-yellow, the veins variably fuscous-scaled and usually with a fairly prominent patch of darker scaling at the bases of the veins emergent from the cell. Hindwing shining white, sometimes with an incomplete greyish postmedian band. An elegant and easily recognised species.

Single-brooded; flies in June and July. Can be disturbed from amongst herbage by day, and flies naturally from dusk onwards into the night, when it will come to light.

LARVA Inhabits a coarse, tubular silken web amongst the seed-head of wild carrot (*Daucus carota*), in August and September. Hibernation occurs in a tough, subterranean cocoon in which pupation occurs in the spring.

DISTRIBUTION Very local in rough fields on light soils and grassy cliff-tops where the foodplant grows. At the time of writing, locally common in the East Anglian Breckland and in chalk quarries in the Thames estuary; also in scattered, often temporary colonies along the south coast to Cornwall. Recorded from Sark, 1982 (Peet, pers. comm.), and twice recently from Jersey (Long, pers. comm.).

Sitochroa verticalis (Linnaeus)

Pl.4, fig.31; text fig.7a

IMAGO Wingspan 30–34mm. Forewing light orange-yellow, the costa and markings orange ochreous; first line indistinct, second evenly curved, serrate, subterminal a little broader than second, undulate with weak serrations extending along veins; a weak, narrow median fascia obliquely basad from irregular reniform stigma; orbicular stigma oval or dot-like; frequently there is a pale rectangular patch between the two stigmata; fringe ochreous white, glossy, with a fine dark subterminal line. Hindwing pale yellowish white with elongate discal spot at about one-quarter, grey cross-lines, the postmedian with a V-shaped incision above middle, the subterminal finely serrate, thickening towards costa; fine dark terminal line and whitish fringe with slightly darker subbasal line. Smaller on average than *Ostrinia nubilalis* (Hübner) and distinguished from the female of that species (Pl.4, fig.35) by the evenly curved second line on forewing, irregularly kidney-shaped discal spot, pale basal half of hindwing and especially by the strongly and characteristically marked underside (text fig.7a). *Microstega pandalis* (Hübner) (Pl.4, fig.33) is a smaller, more slender, greyer insect with differently shaped cross-lines; *M. hyalinalis* (Hübner) (Pl.4, fig.34) is a similar colour to *Sitochroa verticalis*, but the cross-lines are darker and more clearly defined, and second line is shaped like a question mark; the undersides of both these species lack the strongly contrasting blackish fuscous markings found in *S. verticalis*.

Flies in June and July, sometimes with a partial second generation in August and September. Easily disturbed by day from amongst clover and other rough pasture; flies at night and comes to light, and is known to migrate.

LARVA Feeds in late summer on a variety of herbaceous plants including creeping thistle (*Cirsium arvense*), goosefoot (*Atriplex* spp.), dock (*Rumex* spp.), perennial wall-rocket (*Diplotaxis tenuifolia*), *Teucrium* spp. and broom (*Sarothamnus scoparius*). It usually reaches full growth in the autumn, overwinters in a hibernaculum and pupates the following spring.

DISTRIBUTION In fields and pastures in southern England south of Lincolnshire and Herefordshire; stated by Beirne (1952) to be widely distributed and rather common, but it seems to be less so at the time of writing and, like several other pyraustines, its

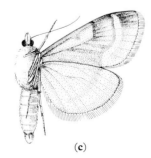

(b)

(a)

Text figure 7

Underside (×2 approx.)

(a) *Sitochroa verticalis* (Linn.)

(b) *Microstega pandalis* (Hb.)

(c) *Ostrinia nubilalis* (Hb.)

(c)

numbers are evidently reinforced from time to time by migration, gradually dwindling between-times. Recorded from the Channel Islands.

Paracorsia repandalis (Denis & Schiffermüller)

Pl.4, fig.32

IMAGO Wingspan 24–28mm. Forewing rather narrow, with straight costa, creamy yellowish white with faint brownish ochreous costal suffusion; three fine, distinct, brown cross-lines, the second of which curves strongly round the discal streak of the same colour, just touching it. Hindwing whitish with clearly marked greyish brown postmedian and subterminal lines. The narrow wings, pale coloration, distinct cross-lines, and especially the discal streak which touches the second line distinguish this species from the other medium-sized pale pyraustines such as *Ebulea crocealis* (Hübner), (Pl.5, fig.12), *q.v.*

LARVA Yellowish white with black dots and a light yellow head. Feeds in spun shoots and flower heads of mullein (*Verbascum* spp.) living gregariously.

DISTRIBUTION Bred from larvae found on dark mullein (*Verbascum nigrum*) near Dawlish, S. Devon, about 1876 (Barrett, 1886), but not reported since. Evidently an extremely rare immigrant from southern Europe.

Microstega pandalis (Hübner)

Pl.4, fig.33, text fig.7b

IMAGO Wingspan 25–29mm. Wings glossy. Forewing very pale whitish yellow, finely irrorate pale grey along costa and termen, and along veins; cross-lines pale grey, rather obscure, first irregularly curved, second oblique, sinuate and weakly serrate, curving round faint, grey, curved discal streak which it does not touch; subterminal line a little broader and darker than the other cross-lines, abruptly indented at approach to tornus; fringe shining whitish. Hindwing whitish, a little paler than forewing, with irregularly sinuate postmedian line and weakly curved subterminal line interrupted by whitish veins; fringe white with faint grey line near base. The markings are weaker than in *Paracorsia repandalis* (Pl.4, fig.32), cross-lines on both wings less sharply defined, and discal streak does not touch second line as it does in that species. See also *Sitochroa verticalis* (Pl.4, fig.31) and female of *Ostrinia nubilalis* (Pl.4, fig.35).

Single-brooded; flies in June. Easily disturbed in daytime from amongst food-plant, and flies gently in vicinity at dusk.

LARVA Feeds on wood sage (*Teucrium scorodonia*), goldenrod (*Solidago virgaurea*) and marjoram (*Origanum vulgare*) in August and September, at first exposed, then in a movable case constructed of leaves of the foodplant or a nearby plant. Hibernates full-fed in the sealed up case, pupating in spring.

DISTRIBUTION Occurs in open parts of woods on light soils where the foodplants grow; widespread in England and Wales and commoner in the south. Widely distributed in the west of Ireland (Pelham-Clinton, pers. comm.).

73

Microstega hyalinalis (Hübner)

Pl.4, fig.34

IMAGO Wingspan 28–35mm. Forewing rather long, arched beyond middle, glossy light yellow, suffused greyish yellow or orange-yellow along costa; cross-lines brown, the first arising from dorsum c. 3mm from base and curving strongly towards base of costa, the second curving strongly round discal spot, often broken on curve, the third following curve of termen and strongly and closely serrate; discal spot oblong, paler in centre; orbicular stigma small, slightly elongate. Hindwing greyish white, postmedian line with a wide curve from middle to costa, subterminal serrate. Cilia very glossy, yellowish white. The characteristic shape of the cross-lines distinguishes *M. hyalinalis* from *Sitochroa verticalis* (Pl.4, fig.31), *q.v.*

Single-brooded; flies in June and July. Hides by day in bushes and other vegetation, from which it is fairly easily disturbed, and flies naturally at dusk and after dark. It comes to light and has been taken at flowers including those of clematis.

LARVA Feeds on common knapweed (*Centaurea nigra*) when young, in a slight web beneath the leaves and making transparent blotches, and after hibernation in a silken gallery near the base of the plant on the young leaves; August to May. Great mullein (*Verbascum thapsus*) is also given as a foodplant (Emmet, 1979). Pupa in June in a white cocoon among leaves of the foodplant.

DISTRIBUTION Local in sheltered places on chalk downs and in clearings in beech-woods where the foodplant grows. Southern counties from Kent to Worcestershire and Dorset; rare in Norfolk (Barrett, 1904). Recorded from Guernsey.

Ostrinia nubilalis (Hübner) European Corn-borer

Pl.4, figs 36(♂),35(♀); text fig.7c

IMAGO Wingspan 29–37mm, females usually the larger. Sexually dimorphic. In male, forewing heavily suffused purplish brown, with exception of narrow cream-coloured band distal to curved, serrate postmedian line, a narrower band basal to first line, and a square or shortly oblong patch between the two stigmata; lines and stigmata themselves very slightly darker than suffusion. Hindwing dull purplish grey with a broad, pale postmedian patch or band. Cilia glossy, purplish grey. In female, the suffusion is weaker, and more or less confined to area between first and second lines, leaving the cross-lines and stigmata more conspicuous. Hindwing less dark, with fairly conspicuous postmedian and subterminal lines, between which lies a broad, pale band. Male is unmistakable; female can be confused with *Sitochroa verticalis* (Pl.4, fig.31), but in that species the second line on forewing is evenly curved, not angled below half-way as in *Ostrinia nubilalis*, and the discal spot is kidney-shaped rather than streaked; in *O. nubilalis*, the basal half of hindwing is usually dark greyish; its underside (text fig.7c) is less strongly marked.

Single-brooded; flies in June and July; seldom seen by day, but comes freely to light and may be encountered flying amongst patches of mugwort after nightfall.

LARVA In this country associated almost entirely with mugwort (*Artemisia vulgaris*), boring in the stem, upwards from an entrance-hole near ground level. The stems snap off easily at the level of the entrance-hole. Full-fed in autumn, overwintering in the burrow and pupating therein the following May. On Continent, a serious pest of maize (*Zea mays*), and recorded also from hop (*Humulus lupulus*), hemp (*Cannabis sativa*), and mallow (*Malva sylvestris*).

DISTRIBUTION Until the 1930s, this species was one of our rarest immigrants. Since then, it has become firmly established on waste ground in the counties bordering the Thames estuary and is taken regularly in London, and appears to be established in some south-coast towns such as Portsmouth and Southampton. Recorded from Ireland at Fountainstown, Co. Cork, for the first time in 1983 (Myers, pers. comm.).

Eurrhypara hortulata (Linnaeus)   Small Magpie

Pl.5, fig.1

IMAGO Wingspan 33–35mm. Head and thorax bright yellow, black-spotted; abdomen black, narrowly banded yellow. The pure white wings, bordered and spotted with black, make this species one of the most familiar and easily recognised pyralids. It varies somewhat in the intensity of the black markings, but extreme forms are uncommon.

 Single-brooded; flies in June and July. Easily disturbed from amongst herbage by day. Flies gently from dusk onwards and comes to light, and is known occasionally to migrate.

LARVA Feeds in August and September on common nettle (*Urtica dioica*), and much less commonly on Labiatae such as white horehound (*Marrubium vulgare*), woundwort (*Stachys* spp.) and mint (*Mentha* spp.), in a rolled leaf or amongst spun leaves. Hibernates, sometimes in numbers, in a transparent cocoon under loose bark or in dead stems of Umbelliferae, in which it pupates in spring.

DISTRIBUTION Common to abundant in most of England south of Yorkshire and Lancashire and throughout Ireland; less common and more local in north England and south Scotland; Isle of Canna (Wormell, 1983). The Channel Islands.

Perinephela lancealis (Denis & Schiffermüller)

Pl.5, figs 2(♂),3(♀)

IMAGO Wingspan 30–34mm. Forewing of male much longer and termen more oblique than in any other British pyraustine with comparable markings, that of female less markedly so. Forewing ground-colour dirty white, heavily suffused and clouded light greyish brown with the exception of a pale area basal to the second line and a pale patch between the stigmata; first and second lines grey-brown, curved, serrate; discal spot narrowly crescent-shaped, orbicular stigma dot-like, both slightly darker than background suffusion. Hindwing of similar colour, with extensive pale area basal to serrate postmedian line. Unmistakable.

Single-brooded; flies in June and July. Hides by day amongst the foodplant and other marsh vegetation and is easily disturbed. Flies at night and comes to light.

LARVA Feeds chiefly on hemp agrimony (*Eupatorium cannabinum*), but recorded also from wood sage (*Teucrium scorodonia*), woundwort (*Stachys* spp.) and common ragwort (*Senecio jacobaea*). Inhabits a leaf with the margins spun downwards, and feeds on leaves, flowers and seed-heads. August to May, hibernating in a silken cocoon in which it eventually pupates.

DISTRIBUTION Most common in rather open, marshy woodland in which hemp agrimony is frequent, but also in more open country and sometimes in woods on drier ground. Widespread but local in England south of Yorkshire, and Wales. Recorded from Co. Cork. Common in Guernsey and Jersey.

Phlyctaenia coronata (Hufnagel)
= *sambucalis* (Denis & Schiffermüller)

Pl.5, fig.4

IMAGO Wingspan 23–26mm. Forewing dark fuscous, very finely irrorate yellowish white; stigmata indistinct, between them a rectangular white spot; another minute white dot near base of wing, and another slightly larger and triangular in shape between interstigmatal patch and tornus; a large, oval white blotch medially in subterminal region, and immediately beyond it traces of a fine subterminal line which curves round it on to costa; distal to this line and between the white blotch and costa is a second, oblique, white blotch incompletely separated into four by three dark veins; a line of small white dots follows course of subterminal line towards tornus. Hindwing similarly coloured, with a large, clear, round white blotch beyond middle of wing; basal to this lies another, more irregular white blotch, and another, separated by a dark vein, lies between the round blotch and costa; beyond these blotches is an irregular subterminal line of small white spots. Fringes banded, yellowish at base, dark fuscous medially and grey distally. This species is similar to *Phlyctaenia stachydalis* (Germar) (Pl.5, fig.10), *q.v.*

Single-brooded; flies in June and July. Rests among elder during the daytime and is easily disturbed from the undersides of the leaves. Flies at dusk and after dark and comes to light.

LARVA Feeds on elder (*Sambucus nigra*), preferring young growth, in a web on underside of a leaf. August and September, hibernating in a silken cocoon amongst fallen leaves or under bark. In the spring, it pupates in this cocoon.

DISTRIBUTION Common amongst elder in England from Durham and Lancashire southwards, and throughout Wales and Ireland; the Channel Islands.

Phlyctaenia perlucidalis (Hübner)

Pl.5, fig.6

IMAGO Wingspan 21–23mm. Forewing and hindwing whitish, when fresh with very weak purplish iridescence; weakly irrorate brownish fuscous towards costa and termen; usual cross-lines weakly defined, brownish; the most conspicuous feature is the dark, lunate discal spot on forewing; terminal neural spots on both wings dark fuscous. The prominent dark discal spot which contrasts with the pale ground-colour and otherwise weak markings preclude confusion with any other species.

Single-brooded; flies in June and July. The first British specimens were taken flying at dusk in Woodwalton Fen, Huntingdonshire (Mere & Bradley, 1957), but most of the others have been captured in the fens at mercury-vapour light. Little is known of its habits in the wild.

LARVA Lhomme (1935) states that the larva feeds on cabbage thistle (*Cirsium oleraceum*), living under a leaf along a vein, from June to August. However, this species of thistle is a rare introduction to the British Isles and is most unlikely to be the British foodplant.

DISTRIBUTION Following the capture of two males in Woodwalton Fen on 24 June 1951, others have occurred in the fens of Huntingdonshire and Cambridgeshire, and also on the coasts of Lincolnshire, Suffolk, Essex and Kent. Recorded as far north as Rudston in south-east Yorkshire in 1983 (Agassiz, 1984).

Phlyctaenia stachydalis (Germar)

Pl.5, fig.10

IMAGO Wingspan 23–25mm. Forewing rather short and blunt, costa evenly arched; fuscous, with very indistinct first and second lines and small dark stigmata; between the stigmata is a rectangular white patch; second line curved very strongly round a white blotch which is finely bisected by a dark vein; immediately distal to the postmedian line is a row of white interneural spots; two other faint pale spots in apex. Hindwing fuscous with darker veins; the most conspicuous markings are two white spots, one in centre of wing, the other half-way between the central spot and base of wing, both bisected by dark veins; distal to the central spot, and curved round it, is a row of small white spots representing the postmedian line, and there is a further row of minute pale dots along termen. Fringes whitish, divided by a fuscous line. Distinguished from *P. coronata* (Pl.5, fig.4) by the blunter forewing, the white spots distal to postmedian line of regular size, not enlarged towards costa as a divided blotch and, in the hindwing, absence of the white blotch between the central one and costa, and the remainder bisected by dark veins.

Single-brooded; flies from June to August. An elusive and secretive species, hiding by day amongst woundwort, often in ditches, difficult to disturb, and skulking close to the patches of foodplant even during its normal nocturnal flight time. Comes to light.

LARVA Feeds in August until October on woundwort (*Stachys* spp.), inhabiting a chamber formed by rolling down the margins of a leaf and binding them with silk. Overwinters in a cocoon in a hollow, dry stem, pupating in spring.

DISTRIBUTION Local and uncommon, probably overlooked, in the southern half of England and south Wales.

77

Mutuuraia terrealis (Treitschke)

Pl.5, fig.5

IMAGO Wingspan 24–28mm. Forewing costa nearly straight to second line, then curved to pointed apex; mauvish grey, closely and evenly irrorate chocolate brown, obscuring the markings; darker towards termen; first line indistinct, very faintly dark-edged basally; second line oblique, widely curved round discal spot above middle, fine, pale, narrowly dark-edged basally; discal spot faint, narrowly crescentic, a little darker than ground colour; orbicular stigma a faint, dark dot; fringe very glossy, grey, with darker line near base. Hindwing ground colour and irroration as in forewing, slightly paler, postmedian line faintly indicated, slightly darker towards termen; fringe whitish, glossy, with darker line near base. The forewing is relatively narrower and longer and more pointed than in either *Opsibotys fuscalis* (Denis & Schiffermüller) (Pl.5, fig.13) or *Udea prunalis* (Denis & Schiffermüller) (Pl.5, fig.17) and is more smoothly marked, less mottled than in either of these species; in *O. fuscalis*, second line is more serrate, and pale, dark-edged postmedian line of hindwing is prominent; in *U. prunalis* the discal spot is broad and conspicuously dark, and the orbicular stigma is also larger and more prominent.

Single-brooded; flies in June and July. Can be disturbed by day from herbage near the foodplant, goldenrod, and flies naturally at dusk. Comes to light later in the night.

LARVA Feeds from July to September on goldenrod (*Solidago virgaurea*) in a loose silken web beneath a leaf or amongst the flowers. Lively, wriggling out of its web when alarmed. Spins a tough cocoon in which it hibernates, pupating in spring.

DISTRIBUTION On rocky coasts and hillsides where the foodplant grows, in western mainland Britain: local in Devon and Cornwall, south and north Wales, northern England and southern Scotland. Recorded from Isles of Rhum and Coll (Wormell, 1983). Irish records need confirmation.

Anania funebris (Ström)
= *octomaculata* (Linnaeus)

Pl.5, figs 7,8; *frontispiece*, fig.8

IMAGO Wingspan 20–23mm. Forewing black with large white subdorsal spot before middle and another in disc beyond middle; sometimes also a small or minute white point between the first and the costa. Hindwing black, with two large white blotches. The presence of the small white point on the forewing is more frequent in western British populations, and in the Burren, Co. Clare, all the white spots are greatly enlarged (fig.8). An unmistakable species.

Apparently single-brooded in south-east England, on the wing in June and July. In the Burren, it flies at the end of May, and a partial second emergence occurs in August. It flies by day. In the Burren, it was seen flying early in the morning, low amongst the foodplant, with the characteristic spinning motion described in the books, difficult to follow and to net; but in the early evening it flew high, fast and straight and could be caught much more easily (pers. obs.).

LARVA Feeds from July until the autumn on the leaves and flowers of goldenrod (*Solidago virgaurea*), inhabiting a slight web under the lower leaves. Hibernates full-grown in an oval yellowish cocoon in which it pupates in spring.

DISTRIBUTION Local in glades and along margins of woods, and on rough hillsides and cliffs inland and on the coast, particularly on limestone, in England and Wales. Recorded in Scotland from Inverness-shire, Argyll, the Isles of Skye and Mull, and from several localities in western Ireland.

Anania verbascalis (Denis & Schiffermüller)

Pl.5, fig.9

IMAGO Wingspan 22–26mm. Forewing deep ochreous yellow, irrorate fuscous to a greater or lesser extent; cross-lines dark fuscous, the first curved on to costa from middle, the second bent at right angles below discal spot and forming a strong curve round it, straightening and thickening at costa; subterminal line wavy or dentate; termen and fringe fuscous; stigmata inconspicuous, but a bright yellow trapezoidal mark between them is a characteristic feature. Hindwing of similar colour, the postmedian line forming a strong zigzag; terminal band fuscous, containing a row of small yellow dots. The species has a characteristic bronze coloration which readily distinguishes it.

Single-brooded; flies in June and July. Rests concealed by day and is not easily disturbed. Flies from dusk onwards and comes to light.

LARVA Feeds on wood sage (*Teucrium scorodonia*) in August and September, first in a tubular gallery and later among leaves which are weakly joined by threads of silk; easily dislodged. Great mullein (*Verbascum thapsus*) is given as an alternative foodplant. Overwinters full-fed in a cocoon and pupates in spring.

DISTRIBUTION Local amongst wood sage growing in open country such as heaths, open conifer plantations and shingle, but not favouring wooded country. Occurs in England south-east of a line between Lincolnshire and Devon, chiefly in the seaboard counties. Frequent in Guernsey (Peet, pers. comm.) and widespread in Jersey (Long, 1967).

Psammotis pulveralis (Hübner)

Pl.5, fig.11

IMAGO Wingspan 23–25mm. Forewing broad, pointed at apex, ochreous irrorate ferruginous; first line indistinct, angled, consisting of a sparse scattering of dark scales; second line fine, straight, bent above middle; subterminal line obscure, slightly darker than ground colour, parallel to termen; discal spot crescentic; veins in terminal region darker than ground colour, reddish brown as other markings; fringe ochreous with two broken ferruginous lines. Hindwing full and broad, light ochreous finely irrorate fuscous brown; postmedian line fine, dark reddish brown, curved parallel to termen; a rather broad brownish subterminal band and fine fuscous terminal line; fringe whitish with broken brown subbasal line. This species is more likely to be mistaken for a hypenine (Noctuidae) at first sight, but the labial palpi are porrected, not ascending, and are hardly longer than the head; the patterned underside is also typically pyraustine in character.

Flies in June and July and with a second generation in August and September, at least on the Continent. Easily disturbed by day from amongst vegetation in marshes.

LARVA Feeds on water mint (*Mentha aquatica*) and gipsywort (*Lycopus europaeus*), living on the undersides of the leaves.

DISTRIBUTION In marshes. Evidently a scarce immigrant which has become temporarily established in the country from time to time, as at Folkestone and in the Isle of Wight in 1869. Nothing has been heard of the species recently. Occurs in marshy places over much of Europe. (See Barrett, 1904; **9**:261.)

Ebulea crocealis (Hübner)

Pl.5, fig.12

IMAGO Wingspan 22–25mm. Forewing varying from yellowish white to yellowish ochreous, the paler specimens weakly suffused ochreous along costa; cross-lines fine, brownish ochreous, the first curving basad to costa, the second forming a strong outward curve round discal spot; discal spot linear, separated from bend of second line by about its own length; orbicular stigma a minute dot, or invisible; a fine reddish brown line along termen; cilia whitish or greyish, glossy. Hindwing greyish white, with very weak indication of postmedian line and discal spot but, as in forewing, with fine reddish brown terminal line; cilia whitish. In the much rarer *Paracorsia repandalis* (Pl.4, fig.32), the forewing is relatively longer, there is a distinct subterminal line, the discal spot touches the second line and the hindwing is more strongly lined. In *Udea lutealis* (Hübner) (Pl.5, fig.15), the cross-lines are more intricately sinuate, the stigmata are larger and clearly outlined, and there is a dark cloud in apex of hindwing. In *U. fulvalis* (Hübner) (Pl.5, fig.16), another rare species, the second line is strongly double-curved near the middle and both stigmata are larger, darker and broader, the orbicular touching the first line.

Single-brooded; flies from late June to August. Rests amongst the foodplant by day and is fairly easily disturbed; flies gently at dusk and later at night and comes to light.

LARVA Feeds from October to June in the heart of a shoot of common fleabane (*Pulicaria dysenterica*) and also ploughman's-spikenard (*Inula conyza*), hibernating when young in a chamber in the turned-down margin or tip of a leaf. Pupa in a loose cocoon among the leaves.

DISTRIBUTION Widespread and locally common in southern England and very local in northern England, southern Scotland, Wales and Ireland, chiefly in marshy areas. Occurs in Guernsey, Jersey and Alderney.

Opsibotys fuscalis (Denis & Schiffermüller)

Pl.5, fig.13

IMAGO Wingspan 20–26mm. Forewing grey with yellowish gloss; cross-lines fine and faint, the first nearly straight, the second serrate, strongly curved round disc, narrowly yellowish-edged distally, slightly thickened at costa and likewise the yellowish edging, forming a moderately prominent pale costal spot; discal spot weak, dark, narrow; orbicular stigma minute, hardly discernible; a very fine, dark terminal line; fringe grey, bisected by a fine dark line. Hindwing grey, with curved, sinuate, dark, pale-edged postmedian line and indication of a pale discal spot; cilia whitish, dark basally. Differences between this and *Mutuuraia terrealis* (Pl.5, fig.5) are given under that species, *q.v.*; *Udea prunalis* (Denis & Schiffermüller) (Pl.5, fig.17) is darker, lacks the yellowish gloss and has wider, darker stigmata.

Single-brooded; flies in June. Easily disturbed during the day from among the foodplants; flies naturally from dusk onwards and comes to light.

LARVA Feeds in flowers and capsules of yellow rattle (*Rhinanthus minor*) and common cow-wheat (*Melampyrum pratense*) in July and August; hibernates in a tough semi-transparent cocoon in which it pupates in the spring.

DISTRIBUTION Locally common in meadows, marshes, open woodland and on northern moorlands where the foodplants grow, from southern England to the Scottish Highlands, the Hebrides, Orkney, Wales and Ireland, and in the Channel Islands.

Nascia cilialis (Hübner)

Pl.5, fig.14

IMAGO Wingspan 24–27mm. Forewing orange yellow, the veins, costa, termen and stigmata ferruginous; cross-lines extremely weak; fringe white with fuscous basal half. Hindwing greyish white. An easily recognised species, though it could conceivably be confused with the scarce immigrant, *Antigastra catalaunalis* (Duponchel) (Pl.5, fig.29), *q.v.*

Single-brooded; flies in June and July. Occasionally disturbed from amongst rank marshland vegetation during the day, but more often seen at light.

LARVA Feeds from August to October on sedges (*Cladium mariscus*, *Carex riparia* and other *Carex* spp.), resting during the day on a leaf and dropping on a thread if disturbed. The full-fed larva hibernates in a papery cocoon in a dead reed-stem, and pupates therein in the spring.

DISTRIBUTION Extremely local in the fens of Cambridgeshire and Huntingdonshire (Gardiner, 1961), with an outlying colony on the Hampshire coast; sometimes fairly common in these restricted localities. Reported from South Thoresby, Lincolnshire, 6 July 1983 (Agassiz, 1984).

Udea lutealis (Hübner)
= *elutalis* auctt.

Pl.5, fig.15

IMAGO Wingspan 23–26mm. Forewing glossy pale whitish ochreous, suffused ochreous along costa; cross-lines fine and faint, ferruginous, the first angulated, second serrulate with deep abrupt sinuation basad below middle; reniform and orbicular stigmata large, a little darker than ground-colour, finely outlined ferruginous; subterminal line weak, broadened in apex to form an oblique, dusky apical streak; termen marked by a fine dark line; fringe whitish, dark-edged. Hindwing whitish, with curved, serrulate, fuscous postmedian line, minute dark discal spot and characteristic dark cloud in apex; fine, broken, fuscous terminal line; cilia yellowish white. An easily recognised species on account of the pale coloration and dark clouding in apex of hindwing.

Single-brooded; flies in July and August. Easily disturbed in daytime from amongst herbage, becoming active before dusk and visiting flowers such as thistles, later coming to light.

LARVA Found in May and June in a web under the lower leaves of a wide variety of herbaceous plants including bramble (*Rubus* spp.), knapweed (*Centaurea* spp.), plantain (*Plantago* spp.), strawberry (*Fragaria vesca*), thistles (*Cirsium* spp.) and mugwort (*Artemisia vulgaris*). Pupa in a folded leaf or between joined leaves in June and July.

DISTRIBUTION Common to abundant on all kinds of waste ground throughout the British Isles. Found in great abundance on low sea-cliffs in the Shetlands (pers. obs.). The Channel Islands.

Udea fulvalis (Hübner)

Pl.5, fig.16

IMAGO Wingspan 24–29mm. Forewing whitish fulvous irrorate fulvous brown; costa and termen fulvous brown; first line oblique from costa to fold, there strongly geniculate and evenly curved to dorsum; second line forms a strong curve from costa to a point dorsal to reniform stigma where it forms a narrow sinus basad before curving on to dorsum; reniform stigma solid, 8-shaped, fulvous brown; orbicular stigma broadly oval, hardly darker than ground-colour, dark-rimmed; veins in terminal region slightly paler than ground-colour; a row of minute, dark interneural terminal dots; fringe whitish, basal half greyish fulvous, outer half divided by a fine dark line. Hindwing light fuscous, slightly darkened towards termen, often with small dark discal dot and traces of postmedian line; fringe whitish, weakly chequered in basal half. Similar in structure, size and markings to *Udea prunalis* (Denis & Schiffermüller) (Pl.5, fig.17), but differs in the dull fulvous brown coloration of the forewing, which is slaty grey in *U. prunalis*, and in the wider submedian sinus of second line (Clutterbuck, 1930; Riley, 1930).

Single-brooded; flies in July and August. Can be disturbed by day and flies at night when it comes to light.

LARVA Hatches in autumn and hibernates when young, apparently without feeding (Fryer, 1933). In spring it feeds on Labiatae, including catmint (*Nepeta cataria*), black horehound (*Ballota nigra*), meadow clary (*Salvia pratensis*), and is full-fed in May or June. Pupa in a cocoon amongst leaves of the foodplant.

DISTRIBUTION An occasional immigrant recorded from Herefordshire and south Dorset, which evidently has become temporarily established from time to time. First recorded from 'the Bournemouth district', 8 August 1927 (Clutterbuck, *loc. cit.*), probably at Parkstone, Dorset, where it was taken in some numbers in the 1930s, and bred. Widespread on the mainland of Europe, and common in Jersey (Long, 1967).

Udea prunalis (Denis & Schiffermüller)
= *nivealis* (Fabricius)

Pl.5, fig.17

IMAGO Wingspan 23–26mm. Forewing fuscous grey, slightly darker on costa and termen, browner towards base; cross-line indistinct, second serrate, darker than ground-colour, paler-edged distally especially on costa, and strongly excavate basad dorsal to reniform stigma; reniform stigma solid, 8-shaped, dark coloured; orbicular stigma smaller, circular, paler. Hindwing fuscous grey, paler towards costa, with weakly defined postmedian line and two small dark discal dots; fringe whitish with dark band through middle. An easily recognised species; see also *U. fulvalis* (Pl.5, fig.16).

Single-brooded; flies in June and July. Hides by day, usually amongst blackthorn, and is easily disturbed. Flies at night and comes to light.

LARVA Hatches in autumn; winter is passed in a white silken hibernaculum at the edge of a leaf. Feeds up in spring until April or May on a wide variety of herbaceous and woody plants, living under the leaves, which it spins together. Recorded foodplants include deadnettle (*Lamium* spp.), woundwort (*Stachys* spp.), dog's mercury (*Mercurialis perennis*), common nettle (*Urtica dioica*), honeysuckle (*Lonicera periclymenum*), bramble (*Rubus* spp.), elder (*Sambucus nigra*) and elm (*Ulmus* spp.). Pupa in a silken cocoon among leaves of the foodplant.

DISTRIBUTION Constantly associated with blackthorn thickets and hedges, and there common to abundant through most of the British Isles, including the Isles of Scilly and the Channel Islands.

Udea decrepitalis (Herrich-Schäffer)

Pl.5, fig.18

IMAGO Wingspan 23–27mm. Forewing very pale greyish ochreous, irrorate brown, browner on costa; first line very indistinct; second line serrate, sinuate; subterminal region with dark suffusion; reniform stigma 8-shaped, brownish, orbicular stigma rounded, faint but darker-edged. Hindwing whitish with brownish grey subterminal border, narrow postmedian line and two small dark discal spots. The pale

83

coloration and dark borders to both fore- and hindwings distinguish this from related species.

Single-brooded; flies from late May to July. Hides by day amongst dense tufts of ferns and other vegetation and moves off inconspicuously if disturbed; sometimes rests on bracken in sunshine with wings spread. No record known of it having been taken at light.

LARVA Feeds in July and August in a slight web on the underside of a frond of narrow buckler-fern (*Dryopteris carthusiana*) and probably other ferns; overwinters in a strong silken cocoon and pupates in spring.

DISTRIBUTION Very local in damp mountain ravines and beside lochs in the Scottish Highlands. Recorded, new to Wales, from near Talybont, Powys, 1978 (Jewess, 1982).

Udea olivalis (Denis & Schiffermüller)

Pl.5, fig.19

IMAGO Wingspan 24–27mm. Forewing greyish ochreous densely irrorate fuscous; cross-lines and stigmata indistinct, the most prominent marking being a trapezoidal white interstigmatal spot; other small white spots are scattered over the wing, especially in the apical region. Hindwing whitish with greyer tornal region, brownish fuscous subterminal band, weak postmedian line and two small dark discal spots, the one nearer the base crescent-shaped. The white patch between the stigmata is the best distinguishing feature.

Single-brooded; flies in June and July. Hides during the day amongst bushes from which it is easily disturbed. Flies at dusk and after dark and comes freely to light.

LARVA Hibernates when young in a turned-down edge of a leaf. Feeds in April, May and early June on many herbaceous plants, including dog's mercury (*Mercurialis perennis*), woundwort (*Stachys* spp.), yellow archangel (*Galeobdolon luteum*), common nettle (*Urtica dioica*), dock (*Rumex* spp.) and hop (*Humulus lupulus*), in spun leaves or the turned-down edge of a leaf. Pupates in a web in a spun leaf.

DISTRIBUTION Occurs in woods, hedgerows and bushy places throughout mainland Britain to southern Scotland, and in Ireland, the Isles of Scilly and the Channel Islands, and is usually common.

Udea uliginosalis (Stephens)
= *alpinalis* sensu Ford

Pl.5, figs 20(♂), 21(♀)

IMAGO Wingspan 25–31mm, females smaller than males. Forewing light greyish brown, virtually unmarked except for indistinctly outlined whitish patch in postmedian region. Hindwing greyish white with darker subterminal band and veins, ground-colour rather darker in female. The rather long, pointed wings which are almost unmarked make this another easy species to identify, but see *U. alpinalis*

(Denis & Schiffermüller) (Pl.5, fig.22).

Single-brooded; flies in July. Easily disturbed from grass and other montane vegetation during the day, when it usually flies off uphill before settling again. Both sexes behave in this way.

LARVA Reported to feed, in June, amongst spun leaves of ragwort (*Senecio* spp.) but, as there is rarely any ragwort where the species occurs in Scotland, presumably in this country on other herbaceous plants (Pelham-Clinton, pers. comm.).

DISTRIBUTION Occurs in colonies in grassy places in the mountains of Scotland above 350m, most frequently in grassy hollows beside streams. Discovered in Orkney in 1973, where it is not uncommon in one locality (Lorimer, 1983). Isle of Rhum (Wormell, 1983).

Udea alpinalis (Denis & Schiffermüller)

Pl.5, fig.22

The species long known in Britain as *alpinalis* has been shown to be *U. uliginosalis*, *q.v.* However, there is a male specimen of true *alpinalis* in the British Museum (Natural History) labelled 'Cumberland [18]83' ex H. J. Turner coll.; genitalia slide: Pyralidae 8570. Perhaps other specimens exist, and the species is possibly resident in parts of upland Britain. Superficially, it differs from *U. uliginosalis* (Pl.5, figs. 20,21) in that the white patch beyond the reniform stigma on the forewing is clearer, whiter and more sharply defined, and the hindwing is whiter with a distinct dark border.

Udea ferrugalis (Hübner)
= *martialis* (Guenée)

Pl.5, fig.23

IMAGO Wingspan 20–24mm. Forewing rusty ochreous, suffused reddish fuscous especially along costa; cross-lines blackish, moderately distinct, the second deeply excavate basad dorsal to reniform stigma; reniform stigma broadly 8-shaped, blackish; orbicular stigma paler, rounded, outlined blackish. Hindwing whitish grey, more or less heavily suffused fuscous, with darker apical suffusion, weak postmedian line and two minute dark discal dots. The rusty forewing with prominent black discal spot is characteristic of this species.

This moth may occur at almost any time of year, but most often in summer and autumn. By day it rests among bushes such as sallows and herbaceous vegetation on waste ground, in gardens and on sea coasts, and is sometimes common in stubble fields. Flies at night and visits flowers such as ivy, over-ripe blackberries, and comes to light.

LARVA Polyphagous on herbaceous plants including woundwort (*Stachys* spp.), hemp agrimony (*Eupatorium cannabinum*), strawberry (*Fragaria* spp.) and burdock (*Arctium* spp.), living at first in a turned-down leaf margin and later among spun leaves. Cocoon made from a partially cut section of leaf, folded and lined with silk. Emergence over a variable period from three weeks to eleven months.

DISTRIBUTION A well-known migrant, numbers of which fluctuate greatly from year to year; recorded from most parts of the British Isles as far north as the Orkneys (Lorimer, 1983).

Mecyna flavalis (Denis & Schiffermüller)
subspecies *flaviculalis* Caradja

Pl.5, fig.24

IMAGO Wingspan 25–29mm. Forewing dull yellow, often suffused grey along costa; cross-lines grey brown, the first oblique, meeting costa at approximately half-distance from base of wing at its junction with dorsum; second line serrulate with strong triangular indentation below middle; subterminal line very faint; reniform stigma oblong, orbicular rounded and with a third smaller rounded spot between it and dorsum, all outlined grey-brown. Hindwing pale yellowish more or less suffused grey, with angulated postmedian line and sometimes two grey discal spots. Both wings have a very narrow, dark terminal line and greyish cilia, producing a sharply pencilled outline to the wings which, together with the presence of a third stigmatal spot in the forewing, makes this species easy to recognise.

Single-brooded; flies in late July and August. Easily disturbed by day from short downland turf, and comes to light at night.

LARVA The biology of this species does not seem to be properly understood. It is stated by Emmet (1979) that the larva occurs in March and April and that the eggs are laid on hedge bedstraw (*Galium mollugo*), field wormwood (*Artemisia campestris*) or small nettle (*Urtica urens*), none of which seem likely to be natural foodplants in Britain.

DISTRIBUTION Extremely local on coastal downland from Kent to Cornwall, and in Wiltshire and Norfolk. A species that has suffered from loss of habitat in recent years.

Mecyna asinalis (Hübner)

Pl.5, fig.25

IMAGO Wingspan 27–34mm. Forewing bluish grey; first line nearly straight, often indistinct and sometimes followed by a blackish triangular subdorsal spot; second line curved, with a deep and abrupt sinuation in middle, often preceded by a large, brownish or dark fuscous trapezoidal subdorsal patch, the line itself sometimes dark-spotted; reniform and orbicular stigmata large, pale, with darker clouding both basally and distally. Hindwing grey with darker postmedian line, strongly sinuate in tornal region. Easily recognised by the long, pointed wings and soft blue-grey coloration.

The moth may be found from May to October, in at least two overlapping broods.

Rather secretive by day, but often encountered at night flying gently on rugged cliffs, or visiting flowers such as heather and ragwort. Comes to light.

LARVA Feeds on wild madder (*Rubia peregrina*), making characteristic large white windows in the leaves, having eaten away the parenchyma from below, leaving the upper epidermis. Flowers and young seeds may also be eaten in summer. Autumn larvae hibernate in rolled dead leaves on the ground. Pupa in a silken cocoon on the ground.

DISTRIBUTION Chiefly western and coastal, following the distribution of wild madder; from Sussex to Cornwall, north and south Devon, Somerset, north and south Wales and Co. Cork in Ireland. Isles of Scilly, common (Agassiz, 1981b); Jersey, not uncommon (Long, 1967).

Nomophila noctuella (Denis & Schiffermüller)　　　Rush Veneer

Pl.5, fig.26; *frontispiece*, fig.9

IMAGO Wingspan 25–34mm. Forewing long and very narrow, with costa straight until near apex, brown clouded with fuscous to a greater or lesser extent; cross-lines weakly defined; a dark median streak at base of wing to first line; the most conspicuous markings are two very large 8-shaped stigmata, the more basal more irregular in shape and evidently composed of fused orbicular and claviform stigmata; usually these markings stand out clearly, but occasionally they hardly contrast with the ground-colour. Hindwing ample, uniform grey-brown with faint discal spot and greyish cilia. Until recently (Munroe, 1973), this was regarded as an absolutely unmistakable species, but it is now known to comprise a species complex of considerable size, one of which, *N. nearctica* Munroe, *q.v.*, has been recorded in Britain.

Moths are most common in summer and autumn, but early immigrants can appear by March. During the day, they are encountered most often on heaths and open country near the sea, and are easily put up. At night, the species comes freely to light.

LARVA Extremely lively and pugnacious, running forwards and backwards with equal facility. Feeds mainly on clover (*Trifolium* spp.), but also on knotgrass (*Polygonum aviculare*) and grasses. Pupa in a tough cocoon amongst grass, or attached to a stone.

DISTRIBUTION One of the best-known of migrant Lepidoptera. Numbers vary from year to year, but often the species reaches extreme abundance by late summer and autumn, and is seldom if ever absent.

Nomophila nearctica Munroe

Pl.5, fig.27

IMAGO Average wingspan a little greater than that of *N. noctuella* (Pl.5, fig.26). In series, ground-colour of forewing distinctly more variegated, with more strongly contrasting maculation, with more extensive pallid dusting, and with a more conspicuous brown shade in the bend of the postmedian line; antemedian line tends to be less conspicuous in *N. nearctica*, and terminal triangular spots are on average less definite and are less often fused distally. Ground-colour of hindwing a little paler and more translucent than in *N. noctuella*, veins darker and more contrasting and terminal infuscation narrower. In male genitalia of *N. nearctica*, cornutus is only about half length of aedeagus, whereas in *N. noctuella* it is nearly as long. In female, structures of the antrum are more complex in *N. nearctica*, and its diverticulum is only about twice as long as broad, whereas in *N. noctuella* it is about five times as long as broad (after Munroe, 1973).

DISTRIBUTION Munroe (*loc.cit.*) found 12 distinct species within the genus *Nomophila*, hitherto supposed to include only the one cosmopolitan species, *N. noctuella*. *N. nearctica* is a widespread North American migratory species. A female specimen discovered in the collection of the late C. W. Mackworth-Praed, labelled 'Southb'r'n/22.viii.19', is considered to have been taken by the late G. H. Heath at Southbourne-on-Sea, Hants, though possibly the label could refer to Southbourne, Sussex (Pelham-Clinton, 1984) – a lesson to us all! This is the only Palaearctic specimen yet known.

Dolicharthria punctalis (Denis & Schiffermüller)

Pl.5, fig.28(♂)

IMAGO Wingspan 20–25mm. Forewing ochreous brown, more or less heavily suffused fuscous; cross-lines indistinct, fuscous; discal spot white, subcrescentic. Hindwing greyish with indistinct dark cross-lines and discal spot. Legs very long; antenna almost reaching apex of forewing. The only other comparable 'leggy' British pyrale is *Synaphe punctalis* (Fabricius) (Pl.6, fig.11), in which the discal spot is dark and the second line is clearly pale-edged distally.

Single-brooded; flies in July and August. Hides by day amongst vegetation close to the sea and becomes active at dusk. Comes to light after nightfall.

LARVA Feeds from September to May on leaves, chiefly dead and decaying, of coastal plants such as bird's-foot trefoil (*Lotus* spp.), clover (*Trifolium* spp.), knapweed (*Centaurea* spp.), plantain (*Plantago* spp.) and even eel-grass (*Zostera* spp.). Pupa in a tough cocoon of silk and vegetable debris, on the ground.

DISTRIBUTION Very local and strictly coastal, on beaches and cliffs by the sea from Sussex to Cornwall and the Isles of Scilly; also Guernsey and Jersey.

Antigastra catalaunalis (Duponchel)

Pl.5, fig.29

IMAGO Wingspan 19–22mm. Forewing very narrow and pointed, termen oblique, pale yellow suffused ferruginous on veins and margins, sometimes so heavily as to obscure ground-colour; cross-lines fine, faint, ferruginous; three small ferruginous subcostal spots, the middle representing the discal spot; fringe white. Hindwing pale yellowish white, ferruginous-tinted especially towards termen; a small blackish discal dot and a larger, round, dark postmedian spot; fringe white. *Nascia cilialis* (Pl.5, fig.14) has somewhat similar coloration but much broader wings.

An extremely scarce vagrant to the southern counties of England; recorded from Hertfordshire, Kent (three times), East Sussex, Isle of Wight, Somerset and Gloucestershire. It is a tropical species which extends into southern Europe.

Maruca testulalis (Geyer)   Mung Moth

Pl.5, fig.30

IMAGO Wingspan 28–34mm. Forewing costa arched, apex pointed, termen slightly concave below apex, warm chocolate brown suffused dark brownish fuscous in median area of wing; a large, narrow, transversely elongate lobed white mark in middle of wing, which is traversed by ferruginous-scaled veins; two smaller white spots between this mark and base of wing; termen very finely white-edged; fringe with narrow blackish subbasal line, brownish fuscous, white towards tornus. Hindwing white with strong chocolate brown border demarcated from basal three-quarters of wing by a strongly sinuate line; the white area is traversed by two or three very fine, serrate, broken lines; two discal spots close to costa; fringe patterned as in forewing.

Two specimens of this cosmopolitan tropical pest species were bred from larvae imported in pods of French beans from Malawi in April 1967 at East Malling Research Station, Kent (Chalmers-Hunt, 1968). In recent years, three wild-taken specimens have been reported: Wanstead Park, Essex, 1979 (de Worms, 1979); Wimbledon, Surrey, 29 July 1983 (Dacie, 1984a); and Mawnan Smith, Cornwall, 15/16 August 1983 (Foster, 1984). Abroad, it is known as the bean pod borer, and infests flowers and pods of numerous legumes.

Diasemia reticularis (Linnaeus)
= *litterata* (Scopoli)

Pl.5, fig.31

IMAGO Wingspan 18–22mm. Forewing brown, in the disc ochreous, irregularly marbled blackish brown; first line white, sinuate from dorsum to fold, then obsolete; second line white, prominent, parallel to termen from costa to midway, then projecting strongly into tornus; discal spot small but clear, white, and in middle of wing a large triangular white mark traversed by blackish-scaled veins; a short white streak in apex; fringe brown, strongly but irregularly chequered white. Hindwing brown, ochreous in discal area, with a prominent white median fascia strongest towards dorsum, and white postmedian line strongly angled into tornus; fringe whitish with dark line near base, chequered only towards apex. Care is needed to differentiate this species and *Diasemiopsis ramburialis* (Duponchel) (Pl.5, fig.32): size and shape are similar, but *D. reticularis* can always be recognised by the postmedian lines of fore- and hindwing, which curve or bend strongly into tornal region; in *D. ramburialis* both proceed weakly towards dorsum, parallel to termen; furthermore, in *D. ramburialis* there is a well-defined white subbasal line on hindwing and dark areas of all the wings are adorned with small, light brown strigulae, easily visible with a hand lens. In *Hymenia recurvalis* (Fabricius), the pattern in both wings is simpler and the fringe is lined, not chequered; in *Pyrausta nigrata* and *P. cingulata* (Pl.4, figs 26,27), the shorter wings are blacker, the markings simpler and the fringes white.

Probably continuous-brooded in the tropics; specimens have been taken in Britain from May to September. Alert and easily disturbed by day from vegetation in dry pastures; flies at night and comes to light.

LARVA Little seems to be known about the life cycle; the larva is said to feed on oxtongue (*Picris* spp.).

DISTRIBUTION A scarce migrant in this country, nowadays less frequently recorded than *D. ramburialis*, but evidently temporarily established in parts of southern England and south Wales during the last century (Bretherton, 1962).

Diasemiopsis ramburialis (Duponchel)

Pl.5, fig.32

IMAGO Wingspan 17–22mm. Forewing chocolate brown, irregularly mottled darker brown and adorned with small, light brown strigulae visible only with a lens; first line whitish, weakly sinuate, extending from dorsum to near costa; second line whitish, tapering form costa, obsolescent from middle, parallel to termen; a square white spot in middle of disc and a white bar dorsal to it, below which is an obscure whitish patch to dorsum; fringe light brown with broad dark brown basal band, more weakly chequered than in preceding species. Hindwing brown and strigulate as forewing, with conspicuous whitish subbasal and median lines; postmedian line a tapering, narrow, sinuate white triangle from costa to halfway, then more distally a short white bar, parallel to termen; fringe white with broken dark brown subbasal band producing a chequered effect. Similar species: *Diasemia reticularis* (Pl.5, fig.31), *q.v.*

This species appears not to have been recorded by day in Britain; specimens have been taken at sugar and light, June, July, September and October.

LARVA Early stages unknown.

DISTRIBUTION A scarce migrant recorded from time to time along the south coast of England and as far north as Lincolnshire. Unlike *D. reticularis* there is no evidence that it has ever bred in this country (Bretherton, 1962).

Hymenia recurvalis (Fabricius)

Pl.5, fig.33

IMAGO Wingspan 22–24mm. Forewing deep brown with broad white median band which curves parallel to termen, with triangular distal projection in cell; a short subapical white bar from costa, dorsal to which are two white spots. Hindwing deep brown with broad white median bar. The markings are simpler than in either *Diasemia reticularis* or *Diasemiopsis ramburialis*, *q.v.*, in both of which the fringes are chequered.

First recorded in Britain at Chiddingfold, Surrey, 5 September 1951 (Mere, 1952); another was taken at Swanage, Dorset, shortly afterwards. Since then there have been a few other records of this migrant from the tropics, where it is widespread and sometimes a pest of spinach, beet, cotton, maize, soya and other plants.

Pleuroptya ruralis (Scopoli)   Mother of Pearl
Pl.6, fig.1

IMAGO Wingspan 33–37mm. Forewing pale whitish ochreous mottled greyish, especially along costa and termen; first cross-line oblique, sinuate, second strongly curved in middle, serrate from curve to costa; discal spot narrowly quadrate, orbicular stigma smaller, round, both grey, separated by pale quadrate mark. Hindwing less ochreous-tinted, with irregular cross-lines and dark discal spot. Cilia of both wings highly glossy. One of the largest, and probably the most familiar of all the British pyrales.

Single-brooded; flies in July and August. Readily disturbed by day from nettle-patches; flies freely at dusk and later comes to light.

LARVA Feeds from early spring until late June or early July on common nettle (*Urtica dioica*), living in a rolled leaf from which it drops on a thread when disturbed, running out forwards or backwards with equal agility. Pupates in a large silk-lined chamber among leaves of the foodplant.

DISTRIBUTION Found throughout the British Isles as far north as the Caledonian Canal; abundant everywhere in the south but more local in the north of its range. Occasionally recorded from Orkney, where it is probably resident (Lorimer, 1983). The Channel Islands.

Herpetogramma centrostrigalis (Stephens)
Pl.6, fig.2

Described from a specimen said to have been taken in Devon (Stephens, 1834). The holotype, which is the only known specimen, is in the British Museum (Natural History). It is figured in Wood (1839). There are some closely similar north American taxa, among which *H. centrostrigalis* will probably be found.

Herpetogramma aegrotalis (Zeller)
Pl.6, fig.3

A specimen, taken by C. S. Gregson near Bolton, Lancashire, was exhibited by C. G. Barrett at the London Entomological Society in March, 1890 (Tutt, 1890a). A southern African species described by Zeller (1852).

91

Palpita unionalis (Hübner)

Pl.6, fig.4

IMAGO Wingspan 27–31mm. Forewing shining translucent white, costa narrowly edged ochreous throughout its length; a minute black discal point and three other such points sometimes visible close to costal streak between middle and base. Hindwing shining translucent white, also with a minute black discal point. Male with conspicuous anal tuft of long grey and white setae. Unmistakable.

A fairly regular immigrant recorded in June but most often in autumn, especially October. Flies at night and comes to flowers, including buddleia and ivy, and to light.

LARVA Feeds in captivity on white jasmine (*Jasminum officinale*). For details of breeding the species, see Howarth (1950). Known to feed in the wild on olive, and probably other Oleaceae (Munroe, pers. comm.).

DISTRIBUTION An uncommon immigrant reported chiefly from the seaboard counties from Kent to Cornwall; also recorded from the Isles of Scilly, Norfolk, Essex, Glamorgan, Co. Cork, and the Channel Islands.

Diaphania hyalinata (Linnaeus) Melonworm
= *lucernalis* (Hübner)

Pl.6, fig.5

IMAGO Wings white like those of *P. unionalis*, but more pointed, and termen of fore- and hindwing more oblique; a broad chocolate-coloured band extends along the length of costa and termen of forewing, and along termen of hindwing. Thorax also chocolate-brown.

One specimen of *D. hyalinata* in British Museum (Natural History) ex Mason coll., labelled as purchased from Knight, a collector, as British. This could be the specimen from Haworth's collection said by Stephens (1834), who also possessed 'a pair captured in Devonshire, near Plymouth', to have been 'taken near London by Mr Knight.' One of the Plymouth specimens is figured by Wood (1839). The species is a native of America.

Agrotera nemoralis (Scopoli)

Pl.6, fig.6

IMAGO Wingspan 20–24mm. Thorax pale yellow, spotted orange. Forewing rather short and broad; basal area to curved first line yellow, orange-marked; remainder of wing light ochreous brown, shaded purplish beyond first line; second line fine, dark fuscous, strongly twice angulated; discal spot ferruginous; fringe pure white with irregularly spaced blackish blotches. Hindwing whitish grey with two incomplete cross-lines; fringe whitish, clouded with grey. A striking and handsome species, unlike any other.

Flies in May and June, sometimes with a partial second emergence in August. Hides amongst foliage of trees during the day and can be beaten out. Flies at night and comes to light.

LARVA Feeds in late June and July on hornbeam (*Carpinus betulus*), at first on the underside of the leaves, later joining two leaves together and eating small round holes in them. Pupa formed in a folded leaf.

DISTRIBUTION Recorded from hornbeam woods in Middlesex, Sussex and Kent; at one time moderately common in its haunts, but only recent records are from Kent, where it seems to persist at very low density. One recorded from Guernsey in July 1982.

Leucinodes vagans (Tutt)

Pl.6, fig.7

Described by Tutt (1890b) from a specimen said to have been taken at Chepstow, Monmouthshire, by Mr. Mason of Clevedon, which was later illustrated (Tutt, 1893). Tutt supposed the moth had come from South America or the West Indies. The whereabouts of the type is unknown; the genus is African, whence several similar species have been described.

Sceliodes laisalis (Walker)

Pl.6, fig.8

IMAGO Wingspan 20–34mm. Forewing costa arched before pointed apex, and termen slightly concave below apex. Forewing ochreous brown; costa narrowly clouded whitish with two or three triangular whitish patches which project towards middle of wing; first line whitish, rather indistinct, curved obliquely basad to dorsum; second line usually distinguishable only towards dorsum, narrow, white, dark-edged basally; subterminal line white, curved from costa to termen and enclosing dark brown apical region, then extending very weakly towards second line before dorsum; median area of wing with irregular brown blotches and angular white marks. Hindwing white with one or two small black discal spots and weakly irrorate brown border from which several wedge-shaped marks extend inwards and reach broken, brown postmedian line; brown irroration extends along dorsum to base of wing; fringe white with strong fuscous subbasal band.

The first British specimen was taken by E. W. Classey at Hampton, Middlesex, 5/6 September 1973 at mercury-vapour light and is in the British Museum (Natural History). Records at Wimbledon, Surrey, 18 July 1983 (Dacie, 1984), and Luton, Bedfordshire, 30 July 1983 (Webb, 1984) suggest possible immigration. An African species which has hitherto been reported from Europe only very occasionally, possibly as a migrant or imported in the early stages in tomatoes, on which the larva feeds.

93

Subfamily PYRALINAE (12 species)

The Pyralinae, or tabbies, are rather broad-winged, retiring moths which inhabit barns, stables, granaries and haystacks. By day they rest on walls inside buildings, usually in dark corners, and scuttle rather than fly when disturbed. The larvae feed on dry plant-material such as hay, or sometimes grain or dried dung. Some species are less common than formerly, probably because of the change to modern farming methods.

Hypsopygia costalis (Fabricius) Gold Triangle

Pl.6, figs 9,10

IMAGO Wingspan 18–22mm. Forewing rosy purple, thinly irrorate fuscous; first and second lines fine, nearly straight, yellow, enlarged on costa to form conspicuous rounded yellow spots; terminal line and fringe yellow. Hindwing usually slightly rosier than forewing, the two fine yellow cross-lines closer together and not enlarged to form costal spots; terminal line and fringe yellow. Occasionally the wings are heavily suffused fuscous, and very rarely the costal spots are enlarged and united (fig.10). Unmistakable.

Single-brooded; flies in July and August. Rests concealed by day in hedgerows, thatch, in clover-ricks and haystacks; sometimes rests indoors on walls of barns. Flies at night and comes to light, sometimes to sugar.

LARVA Feeds from September to May on stored clover and hay, and in squirrels' dreys; probably also in thatch and dead leaves. Pupates in June in an oval cocoon.

DISTRIBUTION Locally common in England south of Durham and Lancashire, in gardens, farmland and woods. Common in Guernsey (Peet, pers. comm.), but Long (1967) gives only two records from Jersey.

Synaphe punctalis (Fabricius)

= *angustalis* (Denis & Schiffermüller)

Pl.6, figs 11(♂),12(♀)

IMAGO Wingspan 22–27mm. Sexually dimorphic; females paler, narrower-winged and heavier-bodied than males. Forewing ochreous, in male more or less heavily irrorate dull reddish purple or reddish brown; first line brown, strongly bowed, usually weak; second line oblique, slightly sinuate, pale ochreous with dark border basally; weak dark discal dot, and several small whitish dots on costa. Hindwing fuscous with trace of postmedian line. In female, the dark border of postmedian line of forewing is much more conspicuous than the pale component, and hindwing is lighter grey, with darker terminal band. Long-legged, like *Dolicharthria punctalis* (Pl.5, fig.28) but that has a white discal spot, and second line is much more irregular and less conspicuous.

Flies from June to August, apparently in one extended emergence. Male often flies naturally by day and is easily disturbed; female is more sluggish, but may also be found on sunny afternoons. Male also flies at night and comes to light.

LARVA Feeds until June, probably from previous autumn, in silken galleries amongst terrestrial mosses, ejecting piles of frass on to the surface, and pupates in a cocoon amongst the moss.

DISTRIBUTION Locally common to abundant on shingle, sandy commons by the sea, dune-slacks, sheltered hollows on chalk-downs and saltmarshes in the southern seaboard counties of England from Kent to Cornwall; also from Somerset, the Norfolk and Suffolk coasts and Breckland, and the Isles of Scilly. Abundant in the Channel Islands.

Orthopygia glaucinalis (Linnaeus)
Pl.6, fig.13

IMAGO Wingspan 23–31mm. Forewing brownish grey with a coppery gloss, unmarked except for two nearly straight, narrow, yellow cross-lines which broaden at costa and about six minute yellow spots along costa between the two lines. Hindwing dull grey, with two almost parallel, fine, curving whitish cross-lines. In both wings there is a very fine, pale terminal line and weakly banded fringes are slightly paler than ground-colour. In female, the long, protrusive ovipositor is a feature. Unmistakable.

Single-brooded; flies in July and August. Hides by day in haystacks, thatch, bundles of dry sticks and similar places and is rather sluggish. At night it flies gently and comes to light and sugar.

LARVA Found in the spring until May in all sorts of decaying vegetable refuse, haystacks, thatch, witches' brooms, birds' nests, dreys, piles of leaves and mowings. Often larvae of very different sizes may be found together. The firm silken cocoon opens by means of a slit at the anterior end, which allows the moth to escape and closes again, leaving no sign of exit.

DISTRIBUTION Local and often fairly common in England south of Durham and Lancashire; has been taken on a lightship more than 30 miles offshore, and may therefore be an occasional migrant. Isles of Scilly, uncommon (Agassiz, 1981b). Three records from Jersey (Long, 1967).

Pyralis lienigialis (Zeller)
Pl.6, fig.14

IMAGO Wingspan 22–26mm. Forewing ferruginous ochreous, heavily suffused dark purplish fuscous; cross-lines white, the first curved sinuate, second strongly sinuate, both broadening rather suddenly at costa to form more or less triangular marks; area between second line and termen suffused grey. Hindwing fuscous with very fine, indistinct postmedian line and dark mark in tornus; fringe grey with darker line near base. A darker species than *P. farinalis* (Linnaeus) (Pl.6, fig.5), lacking contrast

95

between dark basal and terminal areas of forewing and pale median area; cross-lines finer, more conspicuously widened at costa; hindwing darker, much more uniformly coloured with markings almost invisible.

Flies from late June to September, with peak in August, and a single record in May. Rests by day on walls; most records are from rooms inside houses or in farm buildings. The moth rests like *P. farinalis*, with the tip of the abdomen turned up. Flies at night inside barns and houses and comes to light.

LARVA Early stages unknown.

DISTRIBUTION M. F. V. Corley (1985) has found about 30 British records, all from a small area of the south Midlands. The only recent records are from Wiltshire and Oxfordshire, but in the past it has been taken in Buckinghamshire and Gloucestershire. It is evidently a scarce and very local species, though probably overlooked.

Pyralis farinalis (Linnaeus) Meal Moth
Pl.6, fig.15; *frontispiece*, fig.10

IMAGO Wingspan 22–30mm. Forewing ochreous suffused fuscous towards dorsum; cross-lines white, the first curved, second strongly sinuate; area between first line and base of wing purplish brown; terminal area distal to second line also purplish brown, clouded lilac. Hindwing greyish white, crossed by two white lines, similar in shape to those of forewing but more approximated; termen black-spotted, the three near tornus the largest. Similar species: *P. lienigialis* and *P. pictalis* (Curtis), (Pl.6, figs 14;17), *q.v.*

Flies from June to August; may be found resting on walls of stables, outhouses and mills; flies at night and occasionally comes to light and sugar. Resting posture is characteristic, with wings held flat and extended and abdomen curled up over body.

LARVA Lives in a tough silken gallery interwoven with cereal debris and attached to a solid substratum, sometimes for two years. Feeds on stored grain.

DISTRIBUTION Widespread in Great Britain and Ireland and sometimes common in grainstores, stables, barns and mills; seldom encountered at large. Recorded from Guernsey and Jersey.

Pyralis manihotalis Guenée
Pl.6, fig.16

IMAGO Wingspan 24–37mm, female larger than male. Forewing whitish, darker at base to first line, irrorate fuscous; first line fine, white, flattened W- or V-shaped, darker-edged basally; second line whitish, sinuate, broadly edged basally with fuscous irroration; subterminal line of small dark spots; termen with dark interneural dots; discal spot rather large, rounded, blackish. Hindwing similarly patterned. Duller and more uniformly irrorate fuscous than the other British *Pyralis* species, the irroration creating the appearance of an *Aglossa* species with the pattern of a *Pyralis*.

Found from Africa and Madagascar through India to the Far East and Pacific Islands. Larvae were found in 1943 at Dundee, in an importation of bones and animal hides from India (Beirne, 1952).

Pyralis pictalis (Curtis)   Painted Meal Moth
Pl.6, fig.17

IMAGO Wingspan 15–34mm, female larger than male. Forewing blackish to first line, light biscuit-brown between first and second lines, and reddish brown to fuscous brown in terminal area; first line almost straight, only weakly sinuate, fine, white; second line also weakly sinuate, fine, white, nearly parallel to first; discal spot small, circular, black. Hindwing similarly patterned, black-based, with terminal area paler than that of forewing. Most like *P. farinalis* (Pl.6, fig.15) but distinguished at once by the much straighter cross-lines and blackish base of both fore- and hindwings.

A native of the Far East. A specimen was once taken in London (Beirne, 1952). The larva feeds on stored grain and could possibly turn up in warehouses.

Aglossa caprealis (Hübner)   Small Tabby
Pl.6, fig.18

IMAGO Wingspan 23–27mm. Forewing ochreous, heavily mottled and irrorate blackish fuscous and ferruginous; cross-lines ochreous, inconspicuous, the first oblique, serrate, the second almost reduced to a lunate mark near dorsum; termen with obscure blackish streaks between veins; fringe pale ochreous with broken dark band near base. Hindwing whitish with fine dark terminal line; fringe whitish with obscure darker line near base. All wings glossy. A smaller species than *A. pinguinalis* (Linnaeus) (Pl.6, figs 19,20) with more contrastingly blackish marked forewing and white hindwing, not fuscous brown.

Flies in July and August and rests by day on woodwork in and around barns, outhouses and stables.

LARVA Feeds from August to May, or sometimes for two years, in a silken gallery amongst vegetable debris or wheat stacks, ricks and thatch, preferring damp material. Pupa in a tough whitish cocoon to which pieces of straw and debris are attached.

DISTRIBUTION Local and evidently much rarer than formerly; associated with outbuildings in low-lying farmland in England. Recorded from Guernsey.

Aglossa pinguinalis (Linnaeus)   Large Tabby
Pl.6, figs 19(♂),20(♀)

IMAGO Wingspan 30–40mm. Forewing drab ochreous brown heavily irrorate fuscous; three irregular cross-lines, the outer strongly dentate, irregularly edged buff distally; a broad dark costal spot near subbasal line, a roughly triangular one adjacent to first line and a larger, oblong one basal to second line; weak dark discal spot is sometimes fairly prominent. Hindwing fuscous brown, a little paler than forewing, sometimes showing a weak and incomplete pale postmedian line. The only other species of *Aglossa* to be found in the wild in Great Britain is *A. caprealis* (Pl.6, fig.18), *q.v. A. pinguinalis* is perhaps more likely to be mistaken for the noctuid *Athetis pallustris* (Hübner), but this lacks fuscous irroration, has finer cross-lines, the postmedian being much less strongly dentate and lacking the buff markings distally, and also lacks strong costal spots. *Aglossa pinguinalis* very variable in size; female is as large as the male and possesses long, protrusive ovipositor.

Single-brooded; flies in June and July. Rests by day in dark corners of outhouses and barns and, if disturbed, runs rather than flies to a new retreat. Flies at night, mostly inside these buildings, and is only rarely seen at light.

LARVA Lives in silken galleries amongst chaff and hay refuse in barns, and also in sheep-dung inside neglected buildings inhabited by these animals; larvae of all sizes may be found together, and they frequently live for two years.

DISTRIBUTION Much less common than formerly, though still widely distributed in Great Britain and Ireland; also the Isles of Scilly and the Channel Islands. Owing to its retiring disposition it is probably overlooked.

Aglossa dimidiata (Haworth)   Tea Tabby
Pl.6, fig.21

IMAGO Wingspan 20–32mm. Resembles a small *A. pinguinalis* with rather narrower, more pointed wings; first line acutely angled at middle, the angle often containing a conspicuous dark brown arrowhead mark.

A native of the Far East. One specimen in British Museum (Natural History), ex Mason coll., is labelled 'London Docks'.

Aglossa ocellalis Lederer
Pl.6, fig.22

IMAGO Wingspan 14–23mm. More like *A. caprealis* in having white hindwing, but having more uniformly brownish fuscous forewing with a series of small whitish rings in median area between the angulate-sinuate whitish first and second lines.

An African species which was found in Glasgow in 1943 in a cargo of West African palm-kernels (Beirne, 1952).

Endotricha flammealis (Denis & Schiffermüller)
Pl.6, figs 23,24

IMAGO Wingspan 18–22mm. Forewing ochreous yellow, variably suffused reddish purple in basal and terminal areas, and with greater or lesser amount of fuscous irroration; first line fine, whitish, curved round basal patch; second line weakly sinuate, very fine, whitish, dark-edged basally; several small white costal marks; discal spot rounded, blackish. Hindwing similarly patterned but with cross-lines more approximate and terminal reddish purple band thus wider; discal spot absent. Variable in colour from predominantly orange-yellow, through shades of purplish red to almost black (fig.24), but always easily recognised.

Flies in July and August. Hides by day in low bushes, especially oak and gorse, often in bracken, and is easily disturbed. Normal flight-time is from dusk onwards, and it visits flowers and sugar, and comes freely to light.

LARVA When young, inhabits a three- to five-chambered silken cell under a leaf; greater bird's-foot trefoil (*Lotus uliginosus*) is given as a foodplant, and probably several species of deciduous tree and shrub. From late autumn onwards, it feeds on decaying leaves on the ground; full-fed in May, when it pupates in an oval cocoon amongst leaf-litter on the ground.

DISTRIBUTION Local but widespread in the southern half of England and Wales, and recorded from Wigtownshire and also from the Isles of Scilly and the Channel Islands. Occurs in woods, on heaths and waste ground, and is sometimes very common amongst scrub near the sea.

Subfamily GALLERIINAE (7 species)

The Galleriinae are the bee moths or wax moths. Most are associated with the nests of Hymenoptera. The narrow-winged, dull-coloured moths run and scuttle rather than fly, though some species are taken fairly regularly at light. The larvae occur colonially and live in silken galleries, and some are serious pests of hives.

Galleria mellonella (Linnaeus)   Wax Moth

Pl.6, figs 25(♂), 26(♀); *frontispiece*, fig.11

IMAGO Wingspan 30–41mm, females usually larger than males. Sexually dimorphic, termen of forewing in male concave with strong tornal lobe; in female, tornal lobe is hardly discernible. Forewing ochreous brown, second cross-line indicated by short dark streaks; a dark basal streak is usually present, together with a number of dark streaks and spots along dorsum. In female, whole wing often suffused purplish brown, especially towards costa, rendering markings more obscure. Hindwing greyish fuscous in male, paler in female. The most robust British pyralid, and easily recognised.

Flies from June to October, probably in two overlapping broods. Lives in beehives, and probably also the nests of other Hymenoptera, in which it scuttles about with great speed when disturbed. Flies at night around the hives and occasionally comes to light.

LARVA Feeds on honeycomb, which it riddles with its silk-lined galleries, along which it can run with speed if disturbed. When full-grown it constructs a silken cocoon in which the winter larvae hibernate, pupating in spring. Sometimes masses of cocoons occur together.

DISTRIBUTION Much less common than formerly owing to improved bee-keeping techniques, but widespread in England and Ireland, and has been reported from the Inner Hebrides. Taken in Jersey, 1967 and since (Long, pers. comm.).

Achroia grisella (Fabricius)   Lesser Wax Moth

Pl.6, fig.27

IMAGO Wingspan 16–24mm. Head ochreous yellow. Forewing elongate oval, pale shining ochreous fuscous, unmarked. Hindwing a little paler. The characteristic shape, yellow head and absence of wing-markings facilitate identification.

Single-brooded; flies in July and August, with some evidence of a partial second generation, moths having been encountered up to October. Runs actively and rapidly when disturbed inside a beehive, and at dusk flies around the hives, hovering at the entrance like a bee. Very occasionally comes to light.

LARVA Feeds on wax in hives, preferring old wax, but sometimes so abundant as to ruin the hive and cause the bees to desert. Also feeds on dried fruit and dead insects. Very active, thrashing about when disturbed. Pupates in a white silken cocoon.

DISTRIBUTION Common through much of England, Wales and Ireland, and reported from several localities in Scotland; also from the Isles of Scilly, and from Guernsey and Jersey since 1969 (Long, pers. comm.).

Corcyra cephalonica (Stainton)   Rice Moth

Pl.6, fig.30

IMAGO Wingspan 15–22mm. Head whitish. Forewing ellipsoidal, rather pointed, greyish ochreous or light ochreous fuscous, the veins in the median area of the wing from near base to termen dark fuscous streaked, leaving pale zone along costa and a broader one along dorsum; a few small dark marks in termen; fringe concolorous, of long, coarse scales. Hindwing ochreous fuscous, slightly glossy, fringe concolorous, slightly darker towards base. *Achroia grisella* (Pl.6, fig.27) is a similar shape, but has glossier, smoother, unmarked forewing, paler hindwing and bright yellow, not whitish, head.

Flies in July and again in autumn in two overlapping broods. Rests by day on walls and rafters inside buildings, looking like a splinter of wood, sometimes out of doors near infested warehouses. When disturbed, it zigzags away to a new resting place. Flies naturally at night.

LARVA Feeds in July and again through the winter to May on cacao and many other dried foods, including dried fruits, rice, nutmeg, maize, nuts, chocolate and ships' biscuits.

DISTRIBUTION A pest of warehouses, which occurs commonly at times in London, Liverpool and elsewhere.

Aphomia sociella (Linnaeus)   Bee Moth

Pl.6, figs 28(♂),29(♀)

IMAGO Wingspan 25–38mm, females usually larger than males. Sexually dimorphic. *Male*. Forewing creamy buff, fuscous on costa and termen; first and second lines oblique, dentate, blackish; about four blackish elongate subcostal spots evenly spaced from base to near apex, and a row of small blackish interneural spots along termen. *Female*. Forewing browner, more evenly irrorate fuscous, cross-lines and terminal dots more conspicuous; a prominent round black discal spot and another, smaller, half-way between it and first line. Hindwing in both sexes pale fuscous, darker towards apex. Male unmistakable; female bears some resemblance to female *Melissoblaptes zelleri* (Joannis) (Pl.6, figs 31,32), or even *Paralipsa gularis* (Zeller) (Pl.7, fig.2): differences are given under those species.

Single-brooded; flies from June to August. Hides by day in dense vegetation, sometimes resting low down on fences and tree-trunks, and is not easily disturbed. Flies at night and comes to light more often than other Galleriinae.

LARVA Lives in nests of bees (*Bombus* spp.) and wasps (*Vespula* spp.), preferring nests which are above ground. When present, larvae are extremely abundant. They feed at first on old cells and debris in the nest, later attacking the comb and the brood itself, often riddling the nest with silk-lined tunnels. The tough, brown, papery cocoons are often formed in masses; in them, larvae hibernate, pupating in spring.

DISTRIBUTION Widespread and often common through much of Great Britain and Ireland; recorded from the Orkneys and Hebrides. Isles of Scilly, common. Frequent in Guernsey, but not common in Jersey (Long, 1967).

Melissoblaptes zelleri (de Joannis)

Pl.6, figs 33(♂),31,32(♀♀)

IMAGO Wingspan 20–37mm, males smaller than females. Forewing whitish ochreous, sometimes with reddish suffusion towards costa and in female particularly sometimes heavily irrorate fuscous so as to appear nearly black; second line usually weak, angulated and dentate; two small blackish discal spots set in a pale longitudinal streak, spots often absent in male. Hindwing whitish fuscous, slightly opalescent. Narrower-winged than female *Aphomia sociella* (Pl.6, fig.29), less brown, cross-lines less well-defined, discal spots less black and more even in size, and hindwing paler and more glossy. In female *Paralipsa gularis* (Pl.7, fig.2), which is smaller, first cross-line is defined and discal spot larger and more intense than corresponding spot in *Melissoblaptes zelleri*.

Flies June to August. Hidden by day amongst vegetation on sandhills. After dark, on still, warm nights, males may be found on the sand or low herbage, fluttering their wings; both sexes run about on the sand, and the female may make short flights; also reported flying freely in early morning sunshine.

LARVA Feeds on *Brachythecium albicans*, a moss growing on sandhills, inhabiting a vertical silk-lined tube in the sand. Full-fed in May, when it forms a silken cocoon in the tube 3–4cm below the surface (Ford, 1936).

DISTRIBUTION Extremely local on coastal sandhills in Norfolk and Suffolk, east Kent and recorded also from Isle of Wight and Gloucestershire; possibly overlooked elsewhere.

Paralipsa gularis (Zeller) Stored Nut Moth

Pl.7, figs 1(♂), 2(♀)

IMAGO Wingspan 21–32mm, female larger than male. Forewing ochreous buff, weakly irrorate fuscous; cross-lines faint, brownish, the first curved, dentate, pale-edged basally, the second angulate, weakly dentate, pale-edged distally; black discal spot close to angle of second line, much larger and more intense in female; cilia buff. Hindwing whitish, slightly glossy, cilia white. Female distinguished from *Aphomia sociella* by narrower wings and paler coloration; *cf.* also *Melissoblaptes zelleri*. (See Pl.6, figs 29,31,32.)

Flies from May to July in overlapping broods, and may be found resting on the walls of warehouses containing dried fruits, and very occasionally at large in the neighbourhood of such buildings.

LARVA In London, has been found feeding on stored almonds and walnuts from south Europe; also recorded on soya and flax seed. An active larva which wanders restlessly; cocoons, in which the larvae overwinter, are found massed in suitable crannies and crevices in warehouses; emergence has been forced in March (Jacobs, 1935; see also Jacobs, 1933; Wakely, 1933).

DISTRIBUTION Known only from dried-fruit warehouses in London since about 1921; probably elsewhere, if sought.

Arenipses sabella Hampson

Pl.6, fig.3

IMAGO Wingspan 28–46mm. Forewing elongate, light ochreous, finely dusted fuscous especially towards costa; costal margin whitish. Hindwing whitish. The large size and sandy coloration preclude confusion with any other British species; it is more robust than *Melissoblaptes zelleri* (Pl.6, figs 31–33) and quite unmarked.

A female, bred 23 July 1917 from a larva found 3 May 1917 feeding in dates bought in London, was exhibited at the Entomological Society of London. Another female, captured at Canterbury and believed to have come from imported dates, is also mentioned (Durrant, 1919). The former specimen, labelled B. Reid, is now in the British Museum (Natural History) collection. A native of north Africa and the Middle East.

Subfamily PHYCITINAE (58 species)

The Phycitinae comprise just over one-quarter of the British pyralid fauna. They are narrow-winged species; some are brightly coloured, but many are greyish or brownish insects with few markings and are difficult to differentiate. In several species, the males have a characteristic scale-tuft at the base of the antenna, and hence the colloquial name 'knot-horn' which is applied to the group. Many species have two blackish dots, one above the other, in the region of the reniform stigma on the forewing. The majority are strictly nocturnal. The habits of the larvae are diverse, and several are serious pests of warehouses.

Anerastia lotella (Hübner)

Pl.7, fig.4

IMAGO Wingspan 19–27mm. Labial palpi longer than head, straight, porrected. Forewing whitish ochreous, often reddish tinted, weakly to strongly irrorate fuscous except on costal streak, which usually remains conspicuously pale; two small, transversely placed, dark discal dots sometimes present. Hindwing whitish, grey or light fuscous. This species has rather the appearance of a crambid, but more rounded termen and sandy coloration distinguish it. See also *Pima boisduvaliella* (Guenée) (Pl.7, fig.26).

Single-brooded; flies in July. Rests on grass-stems by day and may be tapped out; rests conspicuously on marram and other maritime grasses at night, and comes to light.

LARVA Lives in an irregular tubular silk-lined case on base of stem and among root-stock of grasses such as marram (*Ammophila arenaria*) and sheep's-fescue (*Festuca ovina*) growing in sandy localities. Found in May and June, but probably feeds from previous autumn. Pupates in a silken cocoon in sand near the larval tube.

DISTRIBUTION Common on coastal sandhills in England, Wales, south Scotland and Ireland, and inland on the Breck sands of Norfolk and Suffolk. Two records from Jersey, 1946 and 1964 (Long, pers. comm.).

Cryptoblabes bistriga (Haworth)

Pl.7, fig.5

IMAGO Wingspan 18–20mm. Second joint of antenna of male with sharp, horny apical tooth, easily visible with a lens. Forewing reddish fuscous, darkest in broad median fascia between first and second cross-lines, paler at base and greyer in terminal region; first line broad, oblique, whitish, with thin scattering of reddish scales; second line parallel to termen, serrate, whitish; two very obscure dark discal dots. Hindwing grey. Easily recognised by the broad, reddish fuscous median area of forewing.

Single-brooded; flies in June and July. Hides by day in dense tall bushes and trees and is occasionally disturbed, but usual flight time is at night, when it comes to light.

LARVA Feeds from August to October in a folded leaf of oak, sometimes alder and other deciduous trees, eating holes between the veins. Pupates in autumn in a cocoon amongst detritus on the ground.

DISTRIBUTION Local in oakwoods in England, Wales and Ireland, sometimes found in numbers in just a few trees in a wood. Isles of Scilly, rare (Agassiz, 1981b). Common in Guernsey; taken in Jersey since 1969 (Long, pers. comm.).

Cryptoblabes gnidiella (Millière)

Pl.7, fig.6

IMAGO Wingspan 11–20mm. Forewing whitish or greyish, more or less heavily irrorate black or fuscous; first line whitish extending at right angles from dorsum and curving basad to costa; second line pale, obliquely sinuate dentate from near termen, edged blackish basally; termen with series of blackish interneural dots sometimes enclosed by whitish wedge-shaped markings which project basad; two, sometimes only one, transversely placed, slightly elongate, black discal dots; fringe grey, divided by two broad fuscous bands. Hindwing opalescent whitish with blackish veins and narrow terminal line; fringe whitish with ochreous tint. In *Euzophera bigella* (Zeller) (Pl.8, fig.16), *q.v.*, the cross-lines are not as far apart and much more strongly developed towards dorsum, and first line is nearer to middle of wing.

First recorded in Britain in 1936, when it was bred from a larva in a Jaffa orange purchased at a London street stall (Wakely, 1937). Since then it has been obtained rather frequently from larvae imported in oranges and pomegranates, and has been taken at light. Specimens taken at light have probably emerged from discarded fruit, rather than migrated. A native of the Mediterranean region.

Salebriopsis albicilla (Herrich-Schäffer)

Pl.7, fig.13

IMAGO Wingspan 18–22mm. Sexually dimorphic: in male, head and thickly scaled basal joints of antennae white; female lacks white and specialized antennal scales. Forewing grey, but so heavily irrorate blackish fuscous as to appear dull sooty black. Hindwing fuscous. Smaller and relatively broader-winged than *Pyla fusca* (Haworth) (Pl.7, fig.25), in which the two discal spots are usually visible, but the male is distinguished from that species with certainty by the white head and antennal scapes, which are black in *P. fusca* (Mere, 1965).

Single-brooded; flies in June and July. Flies at night and comes to light.

LARVA Feeds on small-leaved lime (*Tilia cordata*); reported on the Continent from sallow (*Salix* spp.), hazel (*Corylus avellana*) and alder (*Alnus glutinosa*); August to October, living in a rolled leaf.

DISTRIBUTION In Britain known only from the Wye Valley, where it was discovered in 1964, in deep woodland containing small-leaved lime.

Metriostola betulae (Goeze)

Pl.7, fig.24

IMAGO Wingspan 24–27mm. Base of antenna in male slightly thickened. Costa of forewing evenly arched. Forewing blackish fuscous, finely irrorate greyish white and weakly golden lustrous; a rather large tuft of raised black scales before first line, faintly edged whitish basally; first line whitish, conspicuous only towards dorsum, where it is straight and meets dorsum at an angle of c.85°; second line straight and parallel to termen at costa and dorsum, but middle half forms a bulging curve towards termen; discal spot double, the costal component the larger, the two nearly transversely placed; fringe grey with black dividing line. Hindwing translucent fuscous; fringe without gloss, grey with dark dividing line near base. Distinguished from *Pyla fusca* (Pl.7, fig.25) and *Apomyelois bistriatella* (Hulst) (Pl.8, fig.7), *q.v.*, by the presence of scale-tuft preceding first line; a broader-winged insect than *P. fusca*, in which costa is straight to near apex, first line is transversely V-shaped at dorsum and second line makes an irregular zigzag; in *A. bistriatella*, dorsal end of first line is more strongly defined and curves on to dorsum, discal spots are oblique and hindwing more opalescent, with darker border.

Single-brooded; flies in July. Hides by day amongst foliage of birch and usually drops to ground if disturbed. Flies at dusk and after dark and comes to light.

LARVA Feeds in May and June, in a web on upper side of leaf of birch (*Betula* spp.). Pupa in a cocoon amongst spun leaves or in leaf-litter.

DISTRIBUTION Local on sandy heathland with birches, in England south of Durham. Recently found in Dumfriesshire (Pelham-Clinton, pers. comm.). Isles of Scilly, uncommon (Agassiz, 1981b).

Trachonitis cristella (Hübner)

Pl.7, fig.27

IMAGO Wingspan 20–26mm. Forewing whitish, suffused greyish mauve; first line straight, white, edged distally by a broad band of raised blackish red scales; second line straight at costa and dorsum, bowed towards termen in middle two-thirds, whitish, edged basally with a few blackish scales; two small, oblique, dark discal spots set in a pale cloud which extends to costa; a more or less continuous blackish terminal line. Hindwing whitish fuscous, irrorate fuscous towards termen; fringe whitish with dark basal band. A rather dull, poorly-marked species which rather resembles *Numonia advenella* (Zincken) (Pl.8, figs 1,2), but characterized by the conspicuous transverse band of raised dark scales.

Doubtfully British. Barrett had a specimen said to have been taken 'many years ago' on salterns at Portsmouth, Hampshire by H. Moncreaff and originally identified as *Acrobasis tumidana* (= *rubrotibiella* Fischer von Röslerstamm) (Pl.7, fig.33) (Barrett, 1904, **10**:12). In Europe, it has a rather eastern distribution: widespread in S. Germany (Hannemann, 1964), feeding on birch (*Betula* spp.).

Selagia argyrella (Denis & Schiffermüller)

Pl.7, fig.15

IMAGO Wingspan 29–34mm. Forewing highly glossy, ochreous white sparsely dusted fuscous, slightly darker along main veins; two small, transverse discal dots, the upper one the larger; fringe whitish, more ochreous near base. Hindwing light fuscous, fringe whitish ochreous. Another form has the areas between the veins considerably darkened.

Two specimens in British Museum (Natural History): Shoeburyness, 1880 ex Whittle coll. and Adkin coll. ex Mason coll. without further data. The latter is of the dark form.

This is an Asian species with a sparse westerly distribution into Europe.

Microthrix similella (Zincken)

Pl.7, fig.23

IMAGO Wingspan 19–22mm. Male with small thickening at base of antenna. Forewing sooty black, with slight purplish tint in fresh specimens; a conspicuous, straight, slightly oblique, white first line; second line much weaker, often nearly obsolete, irregularly wavy. Hindwing fuscous. Easily identified by the blackish colour with contrasting broad white cross-lines, which reaches almost to costa.

Single-brooded; flies in late June and July. Evidently lives in tree-tops. Comes to light late at night.

LARVA Feeds in July and August in a silk web on leaves of oak (*Quercus robur*). Pupa has been found in a slight web in crevice of bark near base of tree.

DISTRIBUTION Very local in mature oakwoods and in parkland with large oaks. Since the advent of mercury-vapour lights, has been shown to occur in several of the southern counties of England: Hampshire, Sussex, Kent, Surrey, Berkshire and Middlesex.

Pyla fusca (Haworth)

Pl.7, fig.25

IMAGO Wingspan 25–28mm. Base of antenna of male with heavy thickening of coarse scales. Forewing narrow, costa almost straight until just before second line, when it curves into apex. Forewing sooty black, irrorate light grey, the grey irroration defining the cross-lines; first most prominent near dorsum, where it forms a transversely placed V; second forming an irregular zigzag; fringe fuscous with a broken whitish median line; discal spot double, the dorsal component the larger, oblique. Hindwing translucent fuscous; fringe glossy with narrow dark line at base. Distinguishing features of *Metriostola betulae* (Pl.7, fig.24) are given under that species; *Apomyelois bistriatella* (Pl.8, fig.7) is broader-winged, first line is whiter and curves on to dorsum, second line curves in middle towards termen, as in *Metriostola betulae*, and hindwing is more opalescent with more distinct border; dark band in fringe of hindwing is about one-third width of fringe, whereas in *Pyla fusca* it is about one-fifth width.

Single-brooded; flies in June and July. Rests by day amongst heather, especially on burnt twigs, the moths tending to congregate in the burnt areas. Flies at night and comes to light, and to flowers such as ragwort.

LARVA Feeds on heather (*Erica* spp.) from July to the following May, in a silken web. Stated also to feed on bilberry and closely related plants (*Vaccinium* spp.) (Emmet, 1979).

DISTRIBUTION Common on heaths throughout the British Isles, including the Hebrides and Orkneys, Isles of Scilly and the Channel Islands; it is also a wanderer, occasional specimens being taken far away from heathland. It may also breed in heather in gardens.

Phycita roborella (Denis & Schiffermüller)
= *spissicella* (Fabricius)

Pl.7, figs 17,18

IMAGO Wingspan 24–29mm. Antenna of male with strong sinuation at base, filled with large, prominent tuft of scales, absent in female. Forewing reddish fuscous, variably irrorate pale pinkish grey or dark fuscous; in fuscous specimens the irroration may be so dense that most traces of pattern are lost (fig.18); a persistent feature is a triangular dark patch which arises from dorsum at one-third and extends to costa, often preceded by a narrow whitish line; first line oblique, weakly sinuate, lies distal to this patch, separated from it by pale edging; second line parallel to termen from dorsum, then making strong S-curve to costa, fuscous, pale-edged distally; termen with a series of blackish interneural dots; discal spot double, oblique, black. Hindwing translucent fuscous, fringe grey with dark line near base. Female usually less brightly coloured and patterned than male. Easily recognised by the reddish admixture of the forewing, together with the broadly triangular dark patch before first line.

Single-brooded; flies in July and August. Hides by day in lower branches of oak and usually falls to the ground when disturbed. Flies round oak-trees at dusk and into the night, and comes to light, sugar and flowers.

LARVA Feeds up in spring having hibernated, in spun leaves of oak (*Quercus* spp.) from which it can be beaten. Also on pear (*Pyrus* spp.) and crab apple (*Malus sylvestris*). Pupa in soil or leaf litter, or under rotten bark (Hulme, 1968).

DISTRIBUTION Locally common in oakwoods in England and the southern half of Ireland. A few records from Jersey (Long, 1967).

Oncocera semirubella (Scopoli)

Pl.7, figs 7,8

IMAGO Wingspan 26–30mm. Forewing light crimson or pink, sometimes irrorate grey or dark grey; costal stripe white or grey, terminating before apex, or stripe absent (fig.8); dorsum broadly suffused yellow; cross-lines and stigmata absent. Hindwing glossy grey, rosy tinted along termen; fringe paler. Though it flies like a crambid when disturbed, once netted it is unmistakable.

Single-brooded; flies in June and July. Easily disturbed in sunny weather from short turf. Flies from dusk onwards and comes to light and sugar.

LARVA Feeds from August to June on common bird's-foot trefoil (*Lotus corniculatus*) and white clover (*Trifolium repens*). After hibernation it draws together the young shoots in a dense white web; sometimes several larvae inhabit the same web. Pupa in June and July in a dense cocoon on ground or amongst foodplant.

DISTRIBUTION Locally common on chalk and limestone downland and cliffs, from Kent to Somerset and Norfolk. Wanders occasionally.

Pempelia palumbella (Denis & Schiffermüller)

Pl.7, fig.9

IMAGO Wingspan 22–28mm. Male with small but conspicuous scale-tuft at base of antenna. Forewing violet-brown, irrorate greyish white especially towards costa, and with fuscous along veins and dorsum; first line strongly geniculate near costa, pale reddish brown, edged on each side by black scaling; second line with strong sinuation in middle, pale reddish brown, black-edged basally; subterminal line irregularly sinuate, black; a small spot of blackish raised scales on fold immediately preceding first line; discal spot oblique, crescentic or shaped like an inverted T. Recognised by the combination of plumbeous coloration, pale first line and small black scale-tuft.

Single-brooded; flies in July and August. Easily disturbed from amongst heather on sunny days, when it flies away swiftly and erratically, sometimes for a considerable distance. More sluggish in dull weather. Flies naturally from dusk onwards into the night.

LARVA Feeds from August to May on heathers (*Calluna vulgaris* and *Erica* spp.), also on milkwort (*Polygala* spp.) and thyme (*Thymus drucei*), inhabiting a frass-covered web. Pupates in a thick white cocoon in soil.

DISTRIBUTION Often common on heaths in England, southern Scotland, Wales and Ireland. Isles of Scilly, common (Agassiz, 1981b). Also on the Channel Islands, particularly along the cliffs.

Pempelia genistella (Duponchel)

Pl.7, fig.10

IMAGO Wingspan 26–29mm. Male with small scale-tuft at base of antenna. Forewing light brownish ochreous with scattered black scales; veins of disc and termen covered by white scales; first line angled obliquely on to costa, slightly paler than ground-colour, edged with black scales distally; second line sinuate near middle, black-edged basally and distally as a dark wedge in apex; termen with a few irregular semilunar black dots; a small tuft of blackish raised scales preceding first line; discal spot black, elongate basad, sometimes with a faint second dot between it and costa; fringe dull ochreous, faintly lined. Hindwing light greyish ochreous with slight lustre, margin a little darker; fringe whitish with darker basal line. Coloration more ochreous than in related species; black edging of cross-lines usually a prominent feature.

Single-brooded; flies in July. Rests procryptically by day on tip of a broken stem of gorse, whence it can be beaten.

LARVA Feeds on gorse (*Ulex europaeus* and *U. minor*), inhabiting a thick silken web spun in the branches close to the ground; hibernates in a silk tube within the web and feeds up in spring. In fine weather it suns itself on the web, retreating inside when disturbed. Pupates in the web.

DISTRIBUTION Very local on heaths and commons in Sussex, Hampshire, Isle of Wight, Dorset and Wiltshire, mostly near the coast; sometimes common where it occurs. Guernsey, one or two recorded each year since 1981; Jersey, fairly common (Long, 1967).

Pempelia obductella (Zeller)

Pl.7, fig.11

IMAGO Wingspan 23–26mm. Male with tuft of dark scales at base of antenna. Forewing deep purple to orange-brown, the purple coloration of fresh specimens being fugitive; costal streak, terminating before apex, cream or light orange, heavily irrorate white scaling; a broader dorsal streak of similar colour which terminates before tornus, irrorate with a few reddish scales; cross-lines obscure, indicated by white scales, chiefly on veins, the first with a dark dot on vein 3A near dorsum; veins of cell sparsely covered with white scales; discal dots small, faint, dark, surrounded by white irroration; termen with nearly continuous series of black interneural markings; fringe fuscous with darker line. Hindwing glossy, light fuscous, veins darker; fringe whitish with dark line near base. A beautiful and unmistakable species when fresh, but even when worn the forewing retains some suggestion of reddish or purplish which contrasts with the more ochreous costal and dorsal streaks. With a lens, the anterior half of the wing is seen to be delicately irrorate with white.

Flies in July and August. Conceals itself by day but has been beaten from branches and kicked out of vegetation near the foodplant, marjoram. Before dusk, it flies with slow buzzing flight over the foodplant, occasionally settling on it. Flies also at night and comes to light.

LARVA Feeds on marjoram (*Origanum vulgare*) and probably hibernates when young; in early spring may be found in roughly rolled leaves near ground level; later it moves into a terminal shoot, spinning it into a mass and killing it, so that it is overtopped by lateral shoots in a short time. Pupates in a tough, papery cocoon amongst debris on the ground (Huggins, 1929b).

DISTRIBUTION Very rare and local on chalk in Kent where it was shown to be resident by Huggins (1929a), who bred it in 1929. It still persists at very low density.

Pempelia formosa (Haworth)

Pl.7, fig.12

IMAGO Wingspan 20–23 mm. Base of antenna of male with strong scale-tuft. Forewing mottled grey and reddish, redder in dorsal half of wing; first line near middle of wing, pale, strongly toothed basad above dorsum, elsewhere almost obliterated by a broad, irregularly wedge-shaped blackish blotch, broadest at costa; second line faint, irregularly sinuate dentate, edged on each side by irregular blackish crimson scaling; termen with series of shortly crescentic blackish interneural dots; discal spot black, oblique, often contiguous with black median blotch; fringe grey with narrow whitish dividing line. Hindwing grey with ochreous lustre; fringe glossy, grey with dark line near base. This species, which is indeed beautiful, can be recognised by the soft red and grey mottling of the forewing, and prominent median blotch.

Single-brooded; flies from June to August. Hides by day in thick elm hedges, and on the rare occasions when it is disturbed, falls to the ground and conceals itself. Flies from dusk onwards and comes to light.

LARVA Feeds from July to September on elm (*Ulmus* spp.), preferring hedges; it spins a scanty white web on the upper surface of the leaves, but feeds openly by day. Pupates in autumn in earth or rotten wood.

DISTRIBUTION Local amongst hedgerow elms in Hampshire, Isle of Wight, Sussex, Kent, Middlesex, Essex, Suffolk, Norfolk, Hertfordshire and Surrey. Common in Guernsey; four only from Jersey (Long, 1967).

Sciota hostilis (Stephens)

Pl.7, fig.14

IMAGO Wingspan 23–26mm. Male with thickened, coarse-scaled base to antenna. Forewing grey, heavily irrorate fuscous, sometimes with a reddish or orange tint, especially basal to first line; first line pale, angulated on fold and again approaching dorsum, obscure towards costa, broadly edged basally and narrowly distally with blackish fuscous; second line fine, pale, sinuate above middle, narrowly edged dark fuscous; two dark, slightly obliquely placed fuscous discal spots; termen narrowly fuscous, cut by pale ends of veins; fringe greyish fuscous with very fine pale basal line. Hindwing light fuscous, veins slightly darker; termen narrowly darker; fringe greyish fuscous, more clearly divided into darker basal and lighter distal halves than forewing. *Pyla fusca* (Pl.7, fig.25) is a blacker, narrower-winged insect; *Metriostola betulae* (Pl.7, fig.24) has a broad tuft of raised scales on forewing basal to first line; *Apomyelois bistriatella* (Pl.8, fig.7) has more clearly defined, whitish dorsal end to first line, which curves on to dorsum, and more opalescent hindwing with broader, dark border. *Sciota hostilis* is usually a rather dull-coloured, dark, obscurely marked species, though it is variable, and forms in which the base of the forewing is red- or orange-tinted are very easy to recognise.

Single-brooded; flies in June. Rests concealed by day amongst aspen, but sometimes on the trunks when it can be dislodged by jarring the trunk with a stout stick. Flies at night and comes to light.

LARVA Feeds on aspen (*Populus tremula*) from July to September, characteristically attaching a dead leaf to a living one and inhabiting a silken tube inside the chamber so formed. Sometimes more than one larva, each in its own tube, is found in one chamber. Overwinters as a pupa in a tough silken cocoon.

DISTRIBUTION Strictly associated with aspen; scarce and local, recorded from Kent, Essex, Surrey, Worcestershire and Herefordshire.

Hypochalcia ahenella (Denis & Schiffermüller)

Pl.7, fig.22

IMAGO Wingspan 22–32mm. Base of antenna of male slightly thickened and roughened. Labial palpi longer than head, porrect. Forewing greyish ochreous or brownish ochreous with two faint darker cross-lines which converge towards dorsum, sinuate or geniculate; fringe fuscous with blackish line near base. Hindwing fuscous, fringe whitish with dark line near base. A species the very drabness of which precludes confusion.

Single-brooded; flies from June to August. Rests on the ground by day and is easily disturbed, when it flies wildly, often in a circle. Female sometimes rests crossways on a grass-stem near the ground. Natural flight-time is at dusk and during the night, when it comes to light.

LARVA Details of the early stages appear to be unknown, though the larva is said to feed from July to September (Emmet, 1979).

DISTRIBUTION Local on dry, stony ground with sparse vegetation: chalk downs, railway banks, quarries, in southern half of England. Very local in north England, south Scotland and Wales.

Epischnia bankesiella Richardson

Pl.7, fig.16

IMAGO Wingspan 27–30mm. Base of antenna of male hardly thickened. Forewing light grey, veins overlaid blackish; a small, whitish subdorsal spot at one-third often with a small black dot immediately distal to it. Hindwing translucent whitish. The forewing is coloured and marked in a way reminiscent of *Cucullia umbratica* (Linnaeus) (Noctuidae), and no other British phycitid can be confused with it.

Single-brooded; flies in July. Can occasionally be made to fly out from amongst the foodplant in hot weather but the usual time of flight is at dusk and later on at night, when it can be netted as it flies briskly along or boxed resting on the food-plant.

LARVA Feeds on foliage of golden samphire (*Inula crithmoides*) growing amongst limestone rocks, spinning a silken gallery. Hibernates when young, and in spring spins tips of shoots together in a conspicuous manner. Pupates on the ground amongst debris.

DISTRIBUTION Extremely local on limestone cliffs in Dorset, Portland being the best-known locality, and south Wales. Three known colonies in Guernsey.

Dioryctria Zeller

This confused and neglected genus comprises a number of European and north American species of very similar coloration and pattern. Opinions differ over the limits between species; studies of male genitalia reveal the existence of different species groups, but the structures are variable and do not always provide clear differentiation within the groups. Too little is known of the life histories for one to be able to associate a particular mode of life with a particular morph. Moreover, owing to a confusion of names, it is uncertain to which species published details of life histories refer. Shaffer (1966) summarised the position well while showing that *D. sylvestrella* (Ratzeburg) = *splendidella* (Herrich-Schäffer) has no claim to be on the British list. Since that date, indeed since Kloet & Hincks (1972), there have been two significant changes. In the first place, *Dioryctria* f. *mutatella* Fuchs is now generally regarded as being a good species, and *D. schuetzeella* Fuchs has been added to the British list (Chalmers-Hunt & Tweedie, 1982).

There are thus at least three species of *Dioryctria* occurring in Britain, the life cycles of all of which require careful working out. *Dioryctria* larvae have been found in the cones of Scots pine (*Pinus sylvestris*), several larvae inhabiting one cone and emitting pale-coloured frass from holes in the concave side. They have also been found in pith of dead shoots from the previous year's growth; in buds which have been attacked by *Evetria* spp. (Tortricidae); and also in young green shoots, the leaves of which are caused to wither. Some larvae are known to become full-fed in autumn and to hibernate in a flattish hibernaculum on the ground; others have been found to hibernate when small, within the cones. Whether or not these differing habits are characteristic of different species is not known.

Dioryctria abietella (Denis & Schiffermüller)

Pl.7, fig.19; *frontispiece*, fig.12

IMAGO Wingspan 27–33mm. Forewing silvery grey irrorate and mottled fuscous; first line transverse, strongly dentate, silver-grey edged distally with black and basally often by a patch of light brown scales, more numerous towards dorsum; immediately basal to this is a strong, dentate blackish fascia; second line also transverse and strongly dentate, usually with three angles pointing distally and two pointing towards base of wing, edged basally with blackish; subterminal line cloudy, indistinct, grey; discal spot kidney-shaped, usually conspicuously pale, set just distal to a broad blackish median shade; termen black, formed by junction of interneural dots; fringe greyish white, slightly darker in basal half. Hindwing light grey with slightly darker veins and terminal shade; fringe shining white, tinted ochreous near base and with darker dividing line. This species is larger on average than *D. mutatella* Fuchs (Pl.7, fig.20) and the markings are brighter and more contrasting: in particular, the discal spot contrasts more with the ground-colour and the second line is more strongly dentate. In *D. mutatella* this line is more sinuate, with two basally directed dentations and only one, at the middle, directed towards the termen. The genitalia are similar and rather variable, and indicate a close relationship between *D. abietella* and *D. mutatella*, whereas the male organs of *D. sylvestrella*, which may superficially resemble *D. abietella*, show that species to be in a quite different group. The small, broad-winged, brightly marked *D. schuetzeella* Fuchs (Pl.7, fig.21) is best distinguished by the pale mark in the hindwing.

FLIGHT PERIOD and DISTRIBUTION *D. abietella* flies in July and August and is associated with Scots pine (*Pinus sylvestris*). It appears to be widespread in Britain but generally less common than *D. mutatella*. Large specimens, which are taken from time to time away from the vicinity of pine and which are evidently migrants, are *D. abietella* and not *D. mutatella*. Recorded from Guernsey, 1984 (Peet, pers. comm.) and once from Jersey (Long, 1967).

Dioryctria mutatella Fuchs

Pl.7, fig.20

IMAGO Wingspan 24–30mm. Typically smaller and greyer than the preceding species, the markings less contrasting; first line often more oblique than in *D. abietella* (Pl.7, fig.19), and less strongly dentate; light brown scales in basal region of wing absent, and blackish fascia narrower; second line less markedly zigzag, more sinuate, the additional angles close to costa and dorsum, which are typically present in *D. abietella*, are lacking; subterminal line hardly distinguishable in greyer ground-colour of this area of the wing; discal spot grey, hardly contrasting with ground-colour, and median fascia weak or absent. A melanic form occurs in which forewing is entirely sooty black except for a paler patch basal to the (obliterated) first line and slightly paler terminal area; hindwing not appreciably darkened.

FLIGHT PERIOD and DISTRIBUTION *D. mutatella* is on the wing from July to September and is also associated with pine. It is apparently less widespread than *D. abietella* but is the commoner species in East Anglia and perhaps other parts of southern England.

Dioryctria schuetzeella Fuchs

Pl.7, fig.21

IMAGO Wingspan 26–28mm. Forewing relatively shorter and broader than in other

British *Dioryctria*, costa more evenly arched; ground-colour whiter; first line appearing black, strongly dentate, thicker towards costa, edged white basally, the white filling the angle made by the first dentation; basal area mottled ochreous and white with two fine dentate blackish fascia; second line sinuate dentate, white, the strongest dentation near costa, basally black-edged and more narrowly edged fuscous and ochreous along distal edge; discal spot prominent, white; median area of wing mottled white, grey and blackish; termen with a conspicuous black line; fringe grey with broken blackish dividing line. Hindwing light grey with indistinctly outlined but characteristic triangular pale spot near termen, sometimes another on costa near apex and a smaller, rounder one in tornus; fringe white with narrow dark line near base.

FLIGHT PERIOD and DISTRIBUTION The first British specimen was taken at light on 23 July 1980 in Orlestone Forest, Kent, by J. M. Chalmers-Hunt, and in July 1981 it was again taken in this locality and also at Playden, Sussex by M. W. F. Tweedie (Chalmers-Hunt & Tweedie, 1982). In 1982 it was found again, quite commonly, in both these localities, but did not reappear at Playden during 1983.

The species has been bred from Norway spruce (*Picea abies*), a species which occurs both at Orlestone and Playden, from spinnings among the young needles. Such spinnings have been seen at Playden, but so far the larva has not been seen, nor the moth bred, in Britain. It is widespread on the Continent but is not known to migrate, and its origin in south-east England is uncertain.

Pima boisduvaliella (Guenée)

Pl.7, fig.26

IMAGO Wingspan 22–26mm. Base of antenna of male hardly thickened. Forewing brown, becoming brownish ochreous towards dorsum, sometimes reddish-tinged, slightly irrorate blackish; a weakly defined buffish dorsal streak often contains two black dots halfway from base to tornus; clear, narrow, white subcostal streak edged darker suffusion on each side; minute black discal dot. Hindwing grey; fringe whitish. Bears a slight resemblance to *Pleurota bicostella* (Clerck) (Oecophoridae), but that has greyer, more pointed forewing, larger discal spot, and white streak is along costa itself. In *Anerastia lotella* (Pl.7, fig.4), forewing is relatively broader, usually sandier in colour and costal streak is dull creamy white.

Single-brooded; flies from late June to August. Hides among foodplants during the day but flies close to ground in their vicinity on warm, sunny evenings, and may be found resting on vegetation after dark.

LARVA Feeds in pods of sea pea (*Lathyrus japonicus*) and also common bird's-foot trefoil (*Lotus corniculatus*), spiny restharrow (*Ononis spinosa*) and probably other papilionaceous plants, from July to September. Overwinters full-fed in a spherical, sand-covered hibernaculum, but emerges from this to spin a more oval cocoon in which to pupate.

DISTRIBUTION Very local on coastal sandhills and shingle banks from Norfolk to Kent, and in Lancashire.

Nephopterix angustella (Hübner)

Pl.7, fig.31

IMAGO Wingspan 20–25mm. Forewing narrow, greyish ochreous, irrorate reddish ochreous and ashy white, with a few black scales in the disc forming short streaks; first line a little paler than ground-colour, transverse, edged reddish and preceded by a bar of blackish raised scales from dorsum to shortly beyond middle of wing; second line obscure, indicated by a few black streaks towards costa; termen with a row of rather indistinct interneural dots. Hindwing whitish, termen fuscous. Unmistakable.

Flies in June and July, with a partial second emergence in September and October. Hides by day amongst spindle and other bushes, and may sometimes be beaten out. Flies at night and comes to light.

LARVA Lives within berries of spindle (*Euonymus europaeus*), moving from berry to berry, and can be detected by the presence of frass protruding through the entrance-hole, and by a small amount of silk which holds adjacent berries together; July and August, and again in October. When full-grown, the larva enters rotten wood or, in captivity, cork, filling the entrance with gnawings, and spins a papery cocoon; it either pupates at once, the moth emerging in about a fortnight, or overwinters in the cocoon, pupating in spring.

DISTRIBUTION Widespread but local and usually uncommon, from Cambridge and Essex to Herefordshire and Devon; associated with hedgerows and scrubland where the foodplant is established, showing a preference for calcareous soils.

Pempeliella diluta (Haworth)

= *dilutella* auctt.

Pl.7, figs 29,30; text figs 8a,c

IMAGO Wingspan 18–23mm. Antenna of male with conspicuous scale-tuft at base. Forewing ferruginous ochreous mixed crimson, costal half irrorate greyish white to a variable extent; strongly sinuate whitish subbasal line sometimes present; first line oblique, sinuate, whitish, edged reddish or black; second line prominent, complete, weakly sinuate, broadly edged purplish brown on each side; subterminal area sometimes extensively white irrorate; termen with series of black interneural dots; discal spots slightly oblique, black, conspicuous when surroundings are white, not otherwise; fringe fuscous, divided by a fine pale line. Hindwing fuscous or light fuscous, darker terminally; fringe light grey with darker line near base. Variable in size and brightness of forewing markings: specimens from southern England tend to be small and dark, the markings fairly obscure, while those from the west of Ireland are particularly large and bright, with much whitish irroration (fig.30); a large, dark form occurs in the Isle of Man. Easy to identify, however, on account of the narrow, reddish forewing with conspicuous pale second line. Differences between *P. diluta* and *P. ornatella* (Denis &. Schiffermüller) (Pl.7, fig.28) are discussed under that species.

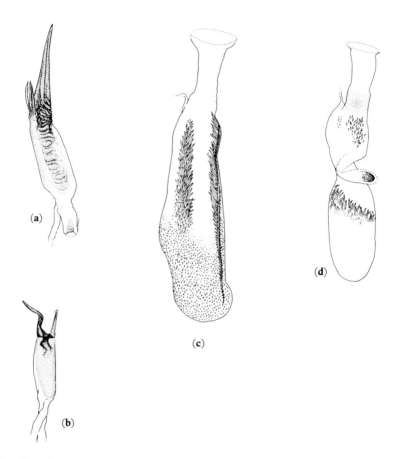

Text figure 8

Pempeliella diluta (Haw.)
(a) male genitalia (aedeagus) (c) female genitalia

P. ornatella (D. & S.)
(b) male genitalia (aedeagus) (d) female genitalia

Single-brooded; flies in July and August. Easily disturbed by day from short calcareous turf; flies in the evening and after dark and comes to light.

LARVA Found in May and June, probably after hibernation, on wild thyme (*Thymus drucei*) in a silken gallery which extends some distance up the plant, and from it on to adjacent stones. Larva inhabits the tip of the gallery, whence it emerges at night to feed. Apparently always associated with yellow ants (*Lasius flavus*). Pupa in a cocoon in soil or amongst leaves or in larval web.

DISTRIBUTION Chiefly coastal on chalk or limestone; also on chalk downs inland. All round the coasts of the British Isles on suitable ground, with the exception of north Scotland. Recorded from the Channel Islands.

Pempeliella ornatella (Denis & Schiffermüller)

Pl.7, fig.28; text figs 8b,d

IMAGO Wingspan 23–27mm. Antenna of male with scale-tuft at base. Forewing brownish ochreous, suffused reddish fuscous along costa and major veins, especially in subterminal area, thus tending to produce a long ochreous stripe between vein 3A and cell, and shorter stripes between veins in outer part of wing; first line consisting of whitish dots or basally directed arrowheads on veins, those on vein 3A and on dorsal rim of cell each with a dark dot in cleft of arrowhead; second line oblique, nearly straight; termen with row of blackish dots, interrupted by white flecks on veins; discal spots small, dark, almost transverse, surrounded by a circlet of white scaling; fringe brownish fuscous, irrorate with white flecks and divided beyond middle by a narrow pale line. Hindwing greyish fuscous, fringe whitish with dark line near base. Tends to be larger than *P. diluta* (Pl.7, figs 29,30), less red; the ochreous streak dorsal to cell appears to be a good separation character as does the straighter, more oblique second line. See also genitalia (text figs 8a–d).

Flies in July and August. Hides by day amongst vegetation and flies when disturbed. Flies at night and comes to light.

LARVA Stated to feed on wild thyme (*Thymus drucei*) and to live in a web amongst the roots (Emmet, 1979).

DISTRIBUTION A local and ill-known species of chalk downs in southern England, chiefly coastal. Burren, Co. Clare, widespread and locally common.

Acrobasis tumidana (Denis & Schiffermüller)

Pl.7, fig.33

IMAGO Wingspan 19–24mm. First joint of antenna of male with horny, scaled apical tooth. Forewing markings similar to those of *A. repandana* (Fabricius) (Pl.7, fig.34), but differs in having a ridge of reddish, raised scales distal to first line, another cluster of such scales at base of wing, and more indistinct, wavy second line; discal dots less distinct.

Single-brooded; flies in July and August. Flies at night and comes to light and sugar.

LARVA Life history apparently undescribed in Britain. Larva feeds in spun leaves of oak (*Quercus* spp.).

DISTRIBUTION A local and little-known species in Britain, recorded in southern England from Kent to Devon, and also from Suffolk, Gloucestershire, Cheshire and north Lancashire.

Acrobasis repandana (Fabricius)
= *tumidella* (Zincken)

Pl.7, fig.34

IMAGO Wingspan 20–25mm. First joint of antenna of male with horny, scaled apical tooth. Forewing grey, more or less densely irrorate reddish fuscous, pinkish brown and fuscous, leaving middle third of wing palest; basal area to first line reddish ochreous, subterminal area reddish fuscous; first line oblique, straight, whitish, edged distally by a wide, fuscous wedge-shaped mark, broadest on costa and fading towards middle of wing; second line oblique, parallel to termen near costa and dorsum, in middle half bulging strongly towards termen, ochreous, edged basally fuscous; discal dots oblique, small, dusky reddish; fringe fuscous with dark line through base. Hindwing fuscous; fringe whitish with dark line near base. Males of *Acrobasis* can be distinguished from those of *Numonia* by the presence of a horny tooth on first joint of antenna, and the first line is straight or at most weakly curved just at costa, pale and dark-edged distally; in *Numonia* species, the first line is angled and never crosses the wing cleanly. *Acrobasis repandana* is less grey than *A. consociella* (Hübner) (Pl.7, fig.32), basal area of forewing is reddish not greyish, and discal spots are smaller; see also *A. tumidana* (Pl.7, fig.33).

Single-brooded; flies in July and August. Hides by day amongst thick masses of oak-foliage and can be disturbed in warm weather. Flies at night and comes to light.

LARVA Feeds on oak (*Quercus* spp.) in May and June, spinning together a small bunch of leaves, usually high up.

DISTRIBUTION Occurs in oakwoods in England south of Durham, and is often common. One record from Jersey, 17 July 1967 (Long, pers. comm.).

Acrobasis consociella (Hübner)

Pl.7, fig.32

IMAGO Wingspan 19–22mm. Antenna of male with horny, scaly tooth at apex of first joint. Forewing greyish white with weak coppery sheen, irrorate dark fuscous, especially in area between first and second lines; base of wing to first line pale greyish; first line oblique, straight, whitish, edged black distally, black area wider at costa; second line fine, sinuate, whitish, narrowly edged brownish fuscous basally; small blackish blotch in apex; fine dark terminal line interrupted by pale veins; discal spots blackish, slightly oblique. Hindwing light fuscous, veins darker; fringe glossy, whitish, with dark line near base. A greyer species than the other British *Acrobasis*, differing particularly in the greyer basal area of the forewing, which is distinctly reddish in the other two species. In some parts of the country, the median and outer parts of the wing are purplish or crimson, but the basal area is always grey. See *A. tumidana* and *A. repandana* (Pl.7, figs 33,34).

Flies in July and August; occasionally there is a small second emergence in October. Hides by day amongst clusters of oak-leaves and falls to the ground if disturbed. Flies from dusk onwards and comes to sugar and light.

LARVA Feeds gregariously in a mass of leaves of oak (*Quercus* spp.) spun together with silk. Hibernates and feeds up in spring. Pupates in a cocoon attached to one of the leaves of the spinning.

DISTRIBUTION Fairly common in oakwoods in England, Wales, southern Scotland and Ireland.

Numonia suavella (Zincken)

Pl.8, fig.4

IMAGO Wingspan 23–25mm. Forewing silvery grey irrorate coppery red or reddish fuscous, especially towards dorsum; first line arched sinuate, meeting dorsum obliquely, whiter on costa, which it meets very obliquely; basal to this line in dorsal half is a triangular area of purplish red irroration, enclosed by a narrow pale line from middle of first line to dorsum, and distal to the first line in costal half is a broader triangle of blackish irroration which gradually fades towards middle of wing; second line whitish, sinuate, at costa extending through an oblong coppery black apical patch; a series of blackish interneural dots in termen; a pair of transversely placed black discal dots, the dorsal the larger; fringe fuscous grey, bisected by a fine pale line. Hindwing light fuscous, veins darker; termen narrowly blackish; fringe whitish with grey line near base. Distinguished from *Acrobasis* species (Pl.7. figs 32–34) by the arched first line and absence, in male, of horny projection on base of antenna; of the British species of *Numonia*, *N. suavella* is the largest and *N. marmorea* (Haworth) (Pl.8, fig.3) is the smallest; *N. suavella* is further distinguished from *N. marmorea* by the fact that in *N. marmorea* the first line reaches dorsum at right angles and is enlarged at that point into a broad whitish patch; furthermore, in *N. marmorea* the two discal spots are nearly always fused. In *N. advenella* (Zincken) (Pl.8, figs 1,2), the line which encloses the reddish dorsal patch basal to the first line starts near costa and not in middle of forewing, the blackish apical patch is absent and the discal dots are oblique, with the dorsal dot nearer the termen.

Flies from mid-July to late August. It seems not to be encountered by day; flies at night and comes to light.

LARVA Feeds in spring on blackthorn (*Prunus spinosa*), spinning a dirty whitish silken gallery close to the branches under the leaves; galleries often rendered conspicuous by presence of frass and pieces of dead leaf which cling to them. Pupa in a greyish cocoon in or near the larval gallery.

DISTRIBUTION Widespread but local and rather uncommon in England south of Herefordshire and Norfolk; associated with stunted blackthorn bushes and thickets in rather open country.

Numonia advenella (Zincken)

Pl.8, figs 1,2

IMAGO Wingspan 19–24mm. Forewing silvery grey irrorate ferruginous or reddish fuscous, especially towards base and termen, basal area usually more fuscous or blackish; first line oblique, twice sinuate between middle and dorsum, grey, stronger towards dorsum; basal to this line is a wedge of coppery red, blackish mixed irroration, enclosed by a fine greyish line which extends from first line near costa to dorsum, and distal to the first line in costal half is a broadly triangular suffusion of similar colour; second line sinuate, silvery grey, edged on each side by reddish suffusion; terminal interneural dots more or less coalescent forming an irregular dark line; discal dots reddish, oblique, the dorsal nearer the termen; fringe fuscous with a darker line. Hindwing grey, veins not darkened; fringe whitish with dark line near base. Rather variable, some specimens being almost uniformly suffused reddish fuscous (fig.2), others showing a considerable amount of silvery ground colour, especially in median area of forewing. Differences from related species are discussed under *N. suavella, q.v.*

Single-brooded; flies in July and August. Hides by day in hawthorn hedges and drops to the ground when disturbed. Flies from dusk onwards and comes to light and sugar.

LARVA Feeds on hawthorn (*Crataegus* spp.) and also rowan (*Sorbus aucuparia*); in spring it spins together a bunch of flowers and buds or young leaves. Pupates in a tough cocoon attached to a dead leaf, or in soil.

DISTRIBUTION Locally common in hawthorn hedges in England, Wales, southern Scotland and Ireland, preferring old, uncut hedges. Recorded from the Isles of Scilly and the Channel Islands.

Numonia marmorea (Haworth)

Pl.8, fig.3

IMAGO Wingspan 18–23mm. Forewing silvery grey irrorate deep crimson or reddish fuscous, especially in basal and terminal areas; first line angled at middle, reaching dorsum at right angles; basal to dorsal half is a triangle of darker crimson coloration which is not, however, enclosed basally by a distinct line; distal to this, the first line expands to form an oblong or triangular white patch of varying width, and in costal half of wing distal to first line is an area of blackish suffusion; second line weakly sinuate, edged on each side with crimson suffusion; an irregular blackish crimson terminal line intersected by greyish veins; discal spots slightly oblique, the dorsal the larger, usually touching. Hindwing greyish fuscous, veins darker; fringe fuscous with darker line near base. *N. marmorea* has relatively shorter, blunter forewing than other British *Numonia*, and costa is straighter; for other differences, see under description (p.119) of *N. suavella* (Pl.8, fig.4).

Flies from June to August with peak in early July. Hides in blackthorn hedges and scrub by day and is not easy to disturb; flies at night and comes to light.

LARVA Feeds in May on blackthorn (*Prunus spinosa*), preferring low stunted bushes. Makes a web on the twigs, which resembles a small piece of sheep's wool. Pupates in the web in a brownish grey cocoon covered with frass and leaf-fragments.

DISTRIBUTION Local in England and Wales, but usually not very common. Good localities include Dungeness, Kent; Eastbourne, Sussex; Portland, Dorset and the Wiltshire Downs; also in the Channel Islands.

Apomyelois bistriatella (Hulst)
subspecies *neophanes* (Durrant)

Pl.8, fig.7

IMAGO Wingspan 18–25mm. Forewing dark fuscous, browner in disc, irrorate whitish towards costa and termen; first line forming a conspicuous whitish curve at dorsum; second line ochreous white, middle half forming a bulging curve towards termen; discal spot double, black, oblique; fringe fuscous, divided by a very narrow white line. Hindwing opalescent greyish with narrow darker border; fringe glossy, whitish, basal one-third dark fuscous. Separated from *Metriostola betulae* (Pl.7, fig.24) and *Pyla fusca* (Pl.7, fig.25) on features discussed under those species. It is rather more variable than either of them, some individuals being almost marbled with grey irroration.

Single-brooded; flies in June and July. Rests by day on old burnt gorse or young birches in heathy localities, head up and thrown back, with wingtips closely appressed to the stem; when dislodged in sunny weather it flies wildly to a new resting-place, but in dull weather it drops to the ground.

LARVA Feeds on the fungus, *Daldinia concentrica*, growing on burnt gorse or young birch. The gallery inside the fungus is silk-lined, and black frass is extruded. Full-fed in October, when it burrows into dead wood, hibernates, and pupates the following spring. Pupae may also be found in the fungus.

DISTRIBUTION Very local in Surrey, Hampshire, Isle of Wight, Dorset and Devon, and found in 1978 on Whixall Moss, Shropshire. Colonies move around as suitable burnt ground becomes available. Jersey, one in 1966, identified by M. Shaffer (Long, pers. comm.).

Ectomyelois ceratoniae (Zeller)   Locust Bean Moth

Pl.8, fig.8

IMAGO Wingspan 19–28mm. Forewing light grey, irrorate fuscous, veins obscurely darker-streaked; cross-lines faint, paler than ground-colour, basally edged darker, the first oblique, strongly indentated near dorsum, second irregularly crenate dentate; two faint, dark, transversely placed discal dots, the one nearer the costa the larger; termen with obscurely darker interneural dots; fringe grey with darker line along base. Hindwing pearly, whitish with darker veins; termen a little darker; fringe whitish. Larger, more robust, with broader wings than *Ephestia kuehniella* Zeller (Pl.8, fig.32), which has more conspicuous blackish edging to cross-lines. Rather variable: ab. *pyralella* Vaughan is whitish, with fuscous irroration confined to apical area of forewing.

Flies from June to September. It is to be found resting inside dried-fruit warehouses, and seems to be taken at light in the wild rather more often than the majority of warehouse pests.

LARVA Feeds inconspicuously within dried fruits, seeds and nuts, such as *Robinia, Ceratonia* beans and dates.

DISTRIBUTION An accidentally introduced species, sometimes common in dried-fruit warehouses; London, Liverpool and probably other ports; sometimes taken at light in the suburbs, probably having originated from local greengrocers' shops.

Mussidia nigrivenella Ragonot

Pl.8, fig.11

IMAGO Wingspan 22–30mm. Forewing grey, irrorate dark fuscous especially along veins; two oblique dark discal flecks sometimes visible; blackish terminal interneural dots present; fringe fuscous grey. Hindwing whitish, darker towards termen especially on veins; fringe whitish with dark dividing line. Somewhat variable in the intensity of ground coloration, but usually recognisable by the dark veins in terminal region of forewing.

A specimen was obtained in a London cocoa warehouse, 18 July 1930 (Potter, 1930). Abroad, the species is a pest of cereals and cacao in West Africa.

Eurhodope cirrigerella (Zincken)

Pl.8, fig.6

IMAGO Wingspan 19–22mm. Forewing rather short for a phycitid, dilated from base, glossy light brownish ochreous, orange-tinged at base, unmarked. Hindwing light fuscous. Although this species resembles no other phycitid, it could perhaps be overlooked among downland crambids; its behaviour, however, is quite different.

Single-brooded; flies from late June to early August. Rests extremely inconspicuously by day under the heads of field scabious, from which it may be disturbed; it then flies off with characteristic buzzing flight. Flies at night and comes to light.

LARVA Feeds in late summer on field scabious (*Knautia arvensis*), normally in the middle flower-head, and up to three or four larvae may occur in each infested head. The larva spins a whitish silken tube in the head, the silk preventing the seeds, on which the larva feeds, from falling. Full-fed in August or September, when it spins a tough, parchment like cocoon in which it hibernates. It may remain a second winter in the cocoon before pupating (Fassnidge, 1928; 1935b).

DISTRIBUTION Nothing seems to have been heard of this species in Britain since 1960 (Goater, 1974). Formerly known from the downs of Wiltshire (Ramsbury, near Marlborough), Berkshire and Hampshire (Farley Mount and Martyr Worthy, whence the last specimen was reported). There is evidence that it demands hot, dry summers, and it must also have suffered severely from loss of habitat.

Myelois cribrella (Hübner)   Thistle Ermine
Pl.8, fig.5

IMAGO Wingspan 27–33mm. Forewing glossy silvery white, sparsely covered with black dots, those nearer base of wing being the larger. Hindwing shining grey; fringe white. *M. cribrella* cannot be mistaken for any other species. The somewhat similar *Yponomeuta* spp. (Yponomeutidae) are much smaller insects.

Single-brooded; flies in July. Rests during the day on thistles, with wings tightly-wrapped round body, and sits more conspicuously on these plants after dark. Occasionally comes to light, but appears not to fly much.

LARVA Feeds in late summer in heads of various thistles (*Cirsium, Carduus, Onopordon*), in the autumn burrowing down into the pith of the stem, where it excavates extensive galleries; several larvae may be found in one stem. Hibernates in the stem when nearly full-grown; in spring, an exit-hole is cut in the dead stem, and pupation occurs in a delicate cocoon near the exit.

DISTRIBUTION Locally common on rough ground, especially on chalk downs and on light, sandy soil, in England south of Lincolnshire and Herefordshire. Isles of Scilly, a few (Agassiz, 1981b); Guernsey, one or two each year (Peet, pers. comm.).

Gymnancyla canella (Denis & Schiffermüller)
Pl.8, figs 9,10

IMAGO Wingspan 22–25mm. Labial palpi long, porrected. Forewing varying from sandy white to pale brick-red and through various shades of whitish grey to dull grey, finely irrorate whitish especially towards termen; paler costal streak; first line a little paler than ground-colour, edged distally by three blackish dots on veins; second line whitish, weakly curved, with blackish fuscous scaling along basal margin, and edged distally with rosy ochreous; discal dots dark, transversely placed, rather far out along wing. Hindwing light grey. This species bears some resemblance to a *Homoeosoma* or *Phycitodes*, but may be distinguished at once by the long, porrected labial palpi which in the other genera are shorter and ascending. Huggins (1927) mentions some interesting aberrational forms bred by him from Kent.

Single-brooded; flies in July and August. May be smoked out of the foodplant by day. Flies at night and comes to light.

LARVA Feeds from August to October on prickly saltwort (*Salsola kali*), burrowing at first into the stems, usually the side-shoots, which are caused to wither at the tips. Later it feeds under a silken web on the unripe seeds, or burrows into the main stem, when it covers the entrance-hole with a few threads of silk. The webs are made conspicuous by the presence of grains of sand trapped among them. Pupation occurs in October in a subterranean cocoon of silk mixed with sand. The pupa may lie over two years (Huggins, *loc.cit.*).

DISTRIBUTION Extremely local and apparently decreasing on sandy coasts just above high tide-mark where the foodplant grows, from Suffolk to Dorset; in 1983, it was found commonly on the Lincolnshire coast.

Zophodia grossulariella (Zincken)

= *convolutella* auctt.

Pl.8, fig.12

IMAGO Wingspan 25–36mm. Forewing whitish, irrorate fuscous especially in dorsal half; first line oblique, angled at middle, curved to dorsum, white, broadly and conspicuously edged distally with black; second line oblique, sinnuate-dentate, often broken, whitish, narrowly edged black basally; a series of rather large, black terminal interneural dots; two transversely placed black discal spots, sometimes united; fringe ochreous white, divided by two or three darker bands. Hindwing whitish, irrorate fuscous especially towards termen; fringe whitish with darker subbasal line. A large, grey phycitid with conspicuously dark-bordered first line.

The larva feeds on unripe gooseberries (*Ribes uva-crispa*). Found in south Europe as far north as Germany and Belgium, and in north America. A specimen was taken near Whitstable, Kent, 30 April 1982 by J. Roche. If it were to become established, it could possibly become a pest (Agassiz, 1983).

Assara terebrella (Zincken)

Pl.8, fig.13

IMAGO Wingspan 22–25mm. Forewing dark reddish fuscous with admixture of white scales; first line whitish, oblique, stepped at vein 3A; second line narrower, sinuate, whitish; a large whitish area in middle of wing from costa encloses two prominent, transverse, dark discal spots; termen reddish fuscous, interrupted by pale veins; fringe grey fuscous with faint darker line. Hindwing light fuscous with whitish fringe containing a darker line. The prominent discal spots, surrounded by pale area, and whitish cross-lines distinguish this species.

Single-brooded; flies from June to August. Flies about spruce at dusk and later comes to light.

LARVA Feeds from September to June in cones of Norway spruce (*Picea abies*), stunting their growth and causing them to fall prematurely. Infested cones lying on the ground are up to *c*.8cm long, black when wet, brittle and easily overlooked amongst other debris on the ground (Fassnidge, 1935a). Pupa in a slight cocoon within the cone.

DISTRIBUTION Local in mature spruce-woods south and east of a line from Norfolk, Gloucestershire and Devon. Isles of Scilly, uncommon (Agassiz, 1981b).

Euzophera cinerosella (Zeller)

Pl.8, fig.17

IMAGO Wingspan 18–25mm. Forewing pale ochreous, irrorate whitish in costal half and in terminal area between the veins; first line dusky, doubly geniculate, very oblique, its costal half forming, together with an oblique dark line basal to second line, a narrowly oval ring which encloses the pale-irrorate cell and two discal spots;

second line sinuate, whitish, dark-edged basally; a shadowy dark mark in apex of wing; discal spots oblique, the one nearer costa closer to apex; fringe concolorous, with a broken darker line through it. Hindwing greyish, veins darker; fringe whitish with darker line. This species is most easily recognised by the narrow dark oval surrounding the pale discal area.

Single-brooded; flies from early June to mid-August. Rests on stems of foodplant near ground during the day, and is seldom seen. Flies at dusk and comes to light later at night.

LARVA Feeds in root-stocks of wormwood (*Artemisia absinthium*) from September to May, forming galleries and chambers in the crown of the root. Pupates in a burrow in pith of an old stem.

DISTRIBUTION Very local amongst wormwood in England south of Derbyshire, and south Wales, chiefly coastal. Well-known localities include Thorpeness, Suffolk, and Portland, Dorset.

Euzophera pinguis (Haworth)

Pl.8, fig.14

IMAGO Wingspan 23–28mm. Forewing light brownish ochreous, sometimes reddish-tinted, irrorate fuscous in costal half; basal half of wing blackish, traversed by ochreous subbasal line, which is indistinct, and sinuate first line; second line sinuate dentate, surrounded by blackish clouding; basal half of fringe ochreous fuscous, outer half white. Hindwing light ochreous fuscous with darker veins; fringe narrowly ochreous basad, with fuscous median line and white edge. The irregular transverse banding of ochreous and black on forewing makes this an easy species to identify.

Single-brooded; flies in July and August. Seldom seen by day, and probably hides in tree-tops, but may sometimes be found freshly emerged on tree-trunks in the evening (Goater, 1974). Flies at night and comes quite freely to light.

LARVA Feeds under living bark of ash (*Fraxinus excelsior*), forming galleries, infesting certain trees which are eventually killed by the larvae, whereupon the colony moves to a neighbouring tree. The presence of black frass thrown out of the entrance hole of the larva betrays its whereabouts. Larvae may be found from September to July, and may live for two years. Pupa in a white silken cocoon under the bark.

DISTRIBUTION Local amongst ash in England south of Yorkshire. Recently noted in Guernsey, and also Jersey (Long, pers. comm.).

Euzophera osseatella (Treitschke)

Pl.8, fig.15

IMAGO Wingspan 22–30mm. Forewing pale whitish buff irrorate light reddish brown; first line irregularly sinuate, strongly curved outwards medially; second line pale, sinuate dentate, strongly edged on both sides dark reddish brown so that the conspicuous feature in this region is a double dark line; basal to this is a rounded, purplish brown cloud; basal area of which containing some long, wedge-shaped dark streaks; a fine, broken blackish line along termen; fringe buff, divided by a darker line. Hindwing whitish ochreous with fine dark terminal line and whitish ochreous fringe divided by a darker line. Distinguished by ochreous buff coloration, strongly dark-edged second line and round dark cloud in middle of wing.

One specimen from East Craigs, Edinburgh, 27 July 1962, bred from larva in a potato imported from Egypt (Whalley, 1963), now in British Museum (Natural History). The museum also contains several specimens bred from Egyptian potatoes in 1963 by Ministry of Agriculture.

Euzophera bigella (Zeller)

Pl.8, fig.16

IMAGO Wingspan 15–20mm. Forewing iron-grey, heavily irrorate with intense black in basal and especially median areas of wing, leaving terminal area paler; first line whitish, angled at fold, broader in dorsal half and obsolescent towards costa; second line whitish, sinuate, wider in dorsal half; fringe grey with broad dark line in basal half. Hindwing light fuscous with fine blackish terminal line; fringe light grey with darker subbasal line. Variable in the amount of black irroration; sometimes there is a fine pale median fascia which broadens towards costa, enclosing as it does a small dark discal spot. Characterised by the grey ground-colour and pale cross-lines which are accentuated towards dorsum. See also *Cryptoblabes gnidiella* (Pl.7, fig.6).

Three specimens have so far been recorded in Britain: the first was bred in 1955 from a larva found in an Italian peach in Edinburgh by E. C. Pelham-Clinton (Shaffer, 1968b); the second was bred from a peach-stone at Hockwold, Norfolk, 15 August 1983 by J. L. Fenn; the third was found resting on a wall in his house at Saffron Walden, Essex, 8 December 1983 (Emmet, 1984).

Nyctegretis lineana (Scopoli)
= *achatinella* (Hübner)

Pl.8, fig.20

IMAGO Wingspan 17–19mm. Forewing light brown, lightly irrorate whitish and fuscous; dorsum more ochreous, weakly irrorate brown; first line straight, very oblique, its point of contact with dorsum twice as far from base of wing as its point of contact with costa, white; second line also straight, nearly parallel to termen, white, darker-shaded basally; discal spot white, black-edged costally. Hindwing light fuscous. The straight, white cross-lines which converge strongly at dorsum, are sufficient to identify this species.

Single-brooded; flies in July and August. Hides amongst dense coastal vegetation by day; flies at dusk over the foodplant, and later at night comes to light, or may be found resting on flowers of viper's bugloss and other plants.

LARVA Inhabits a loose silken tube under common restharrow (*Ononis repens*), and may attack clover (*Trifolium* spp.) and other plants; August to May. Pupates in a silken cocoon attached to the base of a stem of the foodplant.

DISTRIBUTION Very local on flat sandy ground behind coastal sandhills, and on sandy shingle in south-east England from Norfolk to Kent; Thorpeness, Suffolk, is a well-known locality for it.

Ancylosis oblitella (Zeller)

Pl.8, figs 18,19

IMAGO Wingspan 18–22mm. Forewing light grey or greyish ochreous, lightly irrorate fuscous and brown; first line pale, irregularly dentate, not reaching costa, preceded on dorsum by an irregular dark brown or blackish blotch, sometimes faint; second line dentate, slightly concave to line of termen, pale, dark-edged basally; paired discal spots almost transversely placed; a small blackish dot usually present in first notch of first line, on distal side; fringe whitish with faint, darker dividing line. Hindwing opalescent whitish grey, veins darker; fringe glossy, white with very faint, darker line. Variable in the intensity of fuscous suffusion and in sharpness of markings; most easily recognised by the presence of dark blotch on dorsum basal to first line, and pale, dentate cross-lines.

Double-brooded; flies in May and again in July and August. Easily disturbed by day from amongst vegetation on saltmarshes and waste ground; flies at night and comes freely to light.

LARVA Feeds on goosefoot (*Chenopodium* spp.) on saltmarshes and waste ground near the sea. Pupa in a thick silken cocoon above ground. (See Ford, 1957.)

DISTRIBUTION Formerly an extremely scarce immigrant; established on both sides of the Thames estuary since 1956. In 1976, there was a population explosion or large-scale immigration, and moths were reported all along the south coast and inland at least as far as Hertfordshire. Since 1978, it seems to have decreased again.

Homoeosoma sinuella (Fabricius)

Pl.8, fig.21

IMAGO Wingspan 18–23mm. Forewing pale yellowish ochreous, irrorate brownish fuscous along costa, sometimes weakly reddish-tinged; two broad, irregular dark cross-lines, broadest on costa, occasionally broken; frequently also a dark terminal shade. Hindwing grey with darker veins and white, dark-based fringe. An unmistakable species characterised by the ochreous, dark-banded forewing.

Single-brooded; flies from early June to August. Easily disturbed in afternoon sunshine; later in the afternoon it runs with swaggering action (Beirne, 1952) up the stems of grasses, and when disturbed it flies a considerable distance. Flies naturally at dusk in wide zigzags and later comes to light, shuffling about on the sheet in a characteristic manner.

LARVA Lives in root-stocks of plantain (*Plantago* spp.), causing the central leaves to droop in autumn and stunting the growth in spring. Completion of growth is achieved at some time during late autumn, winter or early spring, when the larva makes a silken cocoon in the cavity of the root-stock in which it remains unchanged until May or June.

DISTRIBUTION Locally common on dry, light soils where vegetation is sparse, on cliffs and grassy banks by the sea, waste ground, railway banks, dunes, chalk downs and on the Brecks. In England from Norfolk southwards, and in south Wales. Isles of Scilly (Agassiz, 1981b). Guernsey, very common (Peet, pers. comm.).

Homoeosoma nebulella (Denis & Schiffermüller)

Pl.8, fig.22

IMAGO Wingspan 20–27mm. Forewing very pale ochreous grey with sparse scattering of light brown scales; narrower paler costal stripe tapers to shortly before apex, edged in outer-half along costa itself by a very narrow but characteristic dark shadow; spots representing first line nearly half-way from base to wing-tip, very faint, about four times as long as broad, the uppermost usually absent, the middle furthest from base of wing; two faint discal dots at just over two-thirds distance to apex are slightly elongate and almost exactly transverse; terminal region unmarked save for a faint suggestion of one or two interneural dots; the whole wing with delicate golden lustre. Hindwing whitish hyaline, veins slightly darker, with narrow darker terminal line; fringe white with darker line near base. This is usually larger than *Phycitodes binaevella* (Hübner) (Pl.8, fig.25), though a more delicate-looking insect, and is characterised by the pale coloration, weak markings, and above all by the dark shadow along the outer-half of the costa of the forewing. The genitalia show closest affinity to those of *H. nimbella* (Duponchel) (see text figs 10a–c).

Flies in July, but has been taken in September (Pelham-Clinton, pers. comm.). Seldom seen by day, but flies at dusk and onwards into the night; comes to flowers such as ragwort and thistle, and freely to light.

LARVA Feeds in August and September in flowers and fruiting heads of common ragwort (*Senecio jacobaea*) and spear thistle (*Cirsium vulgare*). Pupates in a loose blackish cocoon in soil or amongst debris on the ground.

DISTRIBUTION Occurs on rough ground on the coast and inland, on chalky or sandy soils; widespread and locally common in England south of Yorkshire, and on the east coast of Ireland.

Homoeosoma nimbella (Duponchel)

Pl.8, fig.23; text figs 9a, 10a,b,g

IMAGO Wingspan 16–21mm. Vein M_2 of forewing weak or absent in this species (text fig.9a): in all the other species of *Homoeosoma* and *Phycitodes* it is present, stalked with vein M_3 (text fig.9b,c). Forewing pale yellowish brown with chalk-

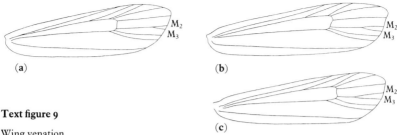

Text figure 9

Wing venation

(a) *Homoeosoma nimbella* (Dup.) M_2 absent, or weak and branching from M_3 beyond two-thirds

(b) *Phycitodes maritima* (Tengst.) M_2 branching from M_3 beyond one-half

(c) *P. saxicola* (Vaugh.) M_2 branching from M_3 before one-half

white subcostal stripe; three small, elongate dark spots represent first line: the middle of these is closest to termen and that nearest costa is closest to base of wing; two dark, transversely placed discal dots are present; subterminal line hardly discernible, weakly edged towards base by a few dark specks. Hindwing light greyish ochreous, glossy; fringe white with faint darker line near base. The least well-understood of the British members of the group (see Pierce, 1937), and evidently confused in the past with *Phycitodes saxicola* (Vaughan) (Pl.8, fig.26) in particular. The neuration of the forewing can be seen easily on examination of the underside with the aid of a binocular microscope, especially using reflected light when the slightly raised veins cast a shadow. In *H. nimbella*, vein M_3 runs nearly parallel to veins CuA_1 and CuA_2, whereas in *Phycitodes* the symmetry is disturbed by the conspicuous fork where vein M_2 branches from M_3. Superficial characters for distinguishing *H. nimbella* are thought to be unreliable, though on the upperside of the forewing the spots are usually browner, less black, and the pale, dark-bordered subterminal line is usually weak or absent in this species. The underside of the forewing is a smoother *café au lait* colour in *H. nimbella*, whereas in *P. saxicola* and *P. maritima* (Tengström) (Pl.8, fig.24) it is whitish, freckled with greyish fuscous. The genitalia (text figs 10a,b,g) show more affinity with those of *H. nebulella* than the species with which it might be superficially confused.

Single-brooded apparently; flies from late May to August. Remains hidden by day, flies at night and comes to light.

LARVA Owing to confusion with related species, it is not possible to state with certainty the details of the life history, though it is assumed that, like the others, the larva feeds in the flowers and seed-heads of Compositae (Roesler, 1973).

DISTRIBUTION Very imperfectly known in Britain. In the British Museum (Natural History) there are specimens from Redruth, Cornwall; Studland, Dorset; Catford, Kent; and a number ex Purdey coll. labelled 'Yarmouth'. Two definite records from Jersey (Long, pers. comm.). Widespread on the Continent, in Norway, Sweden, Denmark, Germany, south and west France.

Text figure 10

Homoeosoma nimbella (Dup.)

(**a**)(**b**) male genitalia
(**g**) female genitalia

Phycitodes maritima (Tengst.)

(**c**)(**d**) male genitalia
(**h**) female genitalia

P. saxicola (Vaugh.)

(**e**)(**f**) male genitalia
(**i**) female genitalia

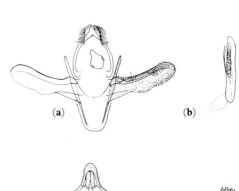

(**a**)　　　　(**b**)

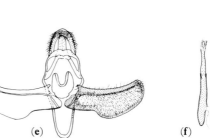

(**c**)　　　　(**d**)

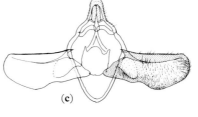

(**e**)　　　　(**f**)

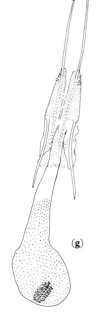

(**g**)

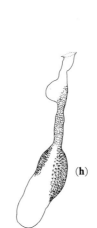

(**h**)

(**i**)

Phycitodes maritima (Tengström)
= carlinella (Heinemann)
= cretacella (Rössler)

Pl.8, fig.24; text figs 9b,10c,d,h

IMAGO Wingspan 18–22mm. Forewing with costa distinctly arched, light straw-coloured mixed greyish, with broad white subcostal streak; three dots representing first line commonly fairly prominent and slightly elongate, the one nearest costa closest to base of wing; the two discal dots are set so that the one closest to costa is very slightly further from termen; second line narrow, pale, dark-edged towards base in costal half. Hindwing light grey, fringe whitish. This species is intermediate in size between *P. binaevella* (Hübner) (Pl.8, fig.25) and *P. saxicola* (Pl.8, fig.26), is shorter-winged than the former and more robust than the latter; its spot pattern resembles that of *P. saxicola*, though the slightly different positioning of the two discal dots does seem to be fairly constant. However, certain identification must rely on a study of the genitalia (see text fig.10), cf. *P. saxicola*. See also *Homoeosoma nimbella* (Pl.8, fig.23).

Moths may be found from May to August and it is probable that there are two broods, though larvae collected during the summer do not seem to emerge until the following year. Flies at night and comes to ragwort flowers and light.

LARVA In flower heads of yarrow (*Achillea millifolium*) and common ragwort (*Senecio jacobaea*), when young in the leaf-axils and upper parts of the stem; sometimes gregarious. Pupates in a cocoon on the ground, in which the autumn larvae overwinter (Huggins, 1929c).

DISTRIBUTION Chiefly coastal in England south of Lancashire and Durham, and known also from the Scottish Highlands (Pelham-Clinton, pers. comm.), but distribution remains imperfectly known.

Phycitodes binaevella (Hübner)

Pl.8, fig.25

IMAGO Wingspan 22–27mm. Forewing chalk-white, irrorate light brown in dorsal half and in terminal area; first line represented by three rather elongate black dots, the dorsal of the three the most elongate and furthest from base of wing; the middle and costal spots are close together, or frequently fused, equidistant from base; two discal spots rather far out on wing, blackish, elongate, transverse; second line whitish, parallel to termen, edged brownish-black along basal edge of costal half, fairly conspicuous; termen with a row of small, dark interneural dots; fringe glossy, whitish. Apart from *Homoeosoma sinuella* and *H. nebulella* (Pl.8, figs 21,22), this is the easiest of the group to identify: it is the most robust and most strongly marked of the remaining species; the three dots on the first line are well-developed and characteristically placed, and it should be unnecessary to refer to the genitalia.

Single-brooded; flies in July. May sometimes be found in the afternoon resting on thistle-heads; flies from dusk onwards, and visits flowers such as thistle and ragwort, and comes to light.

LARVA Feeds in August and September in the seed-heads of spear thistle (*Cirsium vulgare*) and hollows out a large cavity therein. When full-grown, it overwinters in a tough, dull brown cocoon on the ground, and pupates in the spring.

DISTRIBUTION Local but widespread, chiefly on light soils, in England and Ireland.

Phycitodes saxicola (Vaughan)

Pl.8, fig.26; text figs. 9c,10e,f,i

IMAGO Wingspan 14–20mm. Forewing rather narrow, costa not arched, light greyish brown with lighter subcostal streak; three dots which represent first line small and usually round, the one nearest costa, when present, closest to base of wing, the other two more or less equidistant from it; sometimes one or more other dots or short streaks occur along vein 3A; the two small discal dots, which are placed well out along wing, are paired on a line parallel to termen; second line whitish, parallel to termen, narrowly edged brownish towards base in costal half; greyish ochreous, glossy, with faint line near base. Hindwing whitish, fringe glossy, veins and terminal line darker; fringe white, glossy, faintly ochreous at base. This is the smallest species in the group, and in series against *P. maritima* (Pl.8, fig.24) is distinctly less robust, the forewing narrower and with straighter costa, but the antemedian spots do not really seem to provide distinctive separation characters, and the genitalia are the only means of certain identification: in the male of *P. maritima*, the tip of the valve is squared (text fig.10c) and the aedeagus curved and more robust (text fig.10d), whereas in *P. saxicola*, the tip of the valve is rounded (text fig.10e) and the aedeagus is straight and more delicate (text fig.10f); in the female of *P. maritima*, the ductus bursae is bulbed at halfway and the bursa has two very unequal patches of spines (text fig.10h), whereas in *P. saxicola*, the ductus is not bulbed and the two patches of spines on the bursa are of equal size (text fig.10i). See also *Homoeosoma nimbella* (Pl.8, figs 10a,b,g).

Flies from June to August, apparently in one extended brood. It does not seem to be noticed by day, but at night it comes to light.

LARVA British authorities remain uncertain of the life cycle. According to Hannemann (1964), the larva lies in flowers of chamomile (*Anthemis* spp.) and other Compositae, is full-fed in autumn, overwinters in a cocoon and pupates in the spring.

DISTRIBUTION Widespread on coasts of England, Ireland and south-west Scotland; St Kilda (Waterston, 1981). Isles of Scilly, common (Agassiz, 1981b). Taken in several localities in Jersey (Long, 1967).

Plodia interpunctella (Hübner)   Indian Meal Moth

Pl.8, fig.27

IMAGO Wingspan 14–20mm. Forewing ferruginous ochreous, redder or blacker in some individuals according to the extent of dark fuscous irroration; basal area of wing whitish ochreous; first and second lines obscure, dark leaden fuscous; discal spot obscure, pale. Hindwing whitish with darker veins; termen ferruginous; fringe glossy, white with dark line near base. The ferruginous, pale-based forewing is characteristic of this species.

Main emergence is in June and July, but other overlapping broods emerge later in summer and into the autumn. Rests by day with wings folded, resembling a grain of black oats (Beirne, 1952), and becomes active at night.

LARVA Feeds in stored grain, dried fruits, nuts, dried roots and herbs and dried insects. It spins a silken web amongst the food materials which quickly become contaminated with frass. Rate of growth is very variable, from as little as two weeks up to two years to maturity. Pupates in a silken cocoon in a crevice near the food material.

DISTRIBUTION A frequent, often abundant pest of warehouses, first recorded in Britain in 1847. Recorded from Jersey, St. Helier (Long, 1967).

Ephestia Guenée

NOTE. The six species of *Ephestia* which occur in Britain are all very similar and, with the possible exception of *E. kuehniella* Zeller (Pl.8, fig.32), cannot be determined reliably without examination of the genitalia. The superficial characters given below are intended only as a guide and can at best be applied only to typical specimens in bred condition. In caught specimens such markings as exist are more or less abraded. Thus, colour photographs of the imagines are included solely for the sake of completeness and are unlikely, on their own, to be helpful in making accurate identifications. In *E. elutella* (Hübner) and *E. parasitella* Staudinger, veins 3 and 4 of the hindwing are stalked and the cell reaches the middle of the wing; in the other four species, these veins are approximated and the cell falls short of the middle of the wing. (See Hannemann, 1964; Freeman, 1980.)

Ephestia elutella (Hübner)   Cacao Moth

Pl.8, figs 28,29; text figs 11a,12a

IMAGO Wingspan 14–20mm. Head, thorax and forewing brownish ash-grey; first line weakly undulate, oblique, narrow, pale, dark-edged; second line conspicuously pale, oblique, sinuate, dark-edged; two small, transversely placed dark discal spots; dorsum sometimes suffused reddish; hindwing light grey. Genitalia of male without process on dorsal margin of valva (text fig.11a); ductus bursae of female with minute spines scattered in anterior half (text fig.12a). The commonest *Ephestia* species, usually small and pale, with oblique cross-lines and conspicuous pale second line, characters which distinguish it from all but *E. parasitella* (Pl.8, figs 30,31). The ab. *semirufa* Haworth has costal half of forewing grey and dorsal half brown or whitish brown; ab. *roxburghi* Gregson is a melanic form.

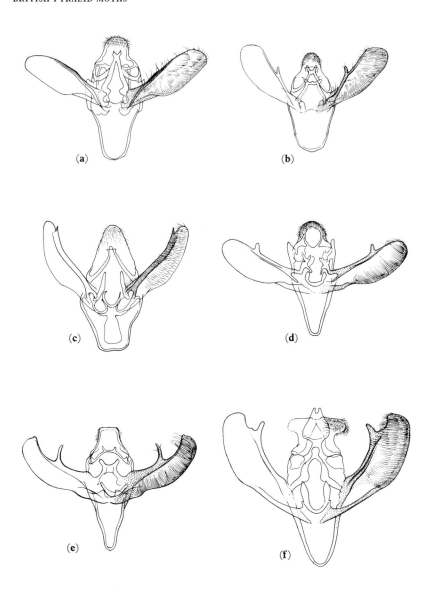

Text figure 11 Male genitalia

(a) *Ephestia elutella* (Hb.) (b) *E. parasitella* Stdgr (c) *E. kuehniella* Zell.
(d) *E. cautella (Walk.)* (e) *E. figulilella* Gregs. (f) *E. calidella* Guen.

Flies from June to October with peak emergence in July and August; hides by day in dark corners of barns and warehouses and flies within such buildings after dark; it has also been taken at light and at flowers.

LARVA Polyphagous on all kinds of stored materials, notably cacao and tobacco, but also on stored cereals on which it can be a serious pest; it also breeds in barns and ricks where it evidently feeds on dry grasses and seeds. Becomes full-grown in about three months though development may take up to a year in unfavourable conditions. When full-fed, the larva wanders before making its cocoon; emergence may take place in as little as ten days, but normally the larva spends the winter in its cocoon before pupating.

DISTRIBUTION The most important insect pest of stored cacao and tobacco in warm and unheated warehouses in Britain, but widespread throughout the country also in farm-buildings, perhaps surviving out of doors in places. Recorded from Guernsey, and common in Jersey (Long, 1967).

Ephestia parasitella Staudinger
subspecies *unicolorella* Staudinger

Pl.8, figs 30,31; text figs 11b,12b

IMAGO Wingspan 14–20mm. Similar to *E. elutella* (Pl.8, figs 28,29) but with forewing often slightly but distinctly broader, suffused crimson reddish in male on base of dorsum and on termen, and with first line more oblique, approaching second line more closely at dorsum than in *E. elutella*; as in that species, the pale, wavy second line is a conspicuous feature. Variable in size and coloration; in one frequent form, the median area of the forewing between the two cross-lines is darkened. The male genitalia differ from those of *E. elutella* in the presence of a short, thumb-like process on the dorsal side of the valva about half-way along its length (text fig.11b); ductus bursae of female dentate just before bursa copulatrix (text fig.12b).

Flies from June to September; has been beaten from ivy, yew and alder and at night it comes to light. Unlike the other British species, it is not a warehouse pest and has been found only out of doors.

LARVA Becomes full-fed in autumn and hibernates in a silken cocoon, pupating in spring. Little is known of its habits in the wild, but it probably feeds on dried plant material, including old, dried berries and dead stems of ivy.

DISTRIBUTION Reputed to be local and rare in the southern counties of England, but probably overlooked and possibly increasing. Reported from Essex, Kent, Sussex, Hampshire, Surrey, Middlesex, Devon, Somerset, Gloucestershire, Herefordshire and Monmouthshire. Jersey, 3 since 1962 (Long, 1967).

Ephestia kuehniella Zeller   Mediterranean Flour Moth

Pl.8, fig.32; text figs 11c,12c

IMAGO Wingspan 20–25mm. Forewing grey, irrorate whitish fuscous in middle and

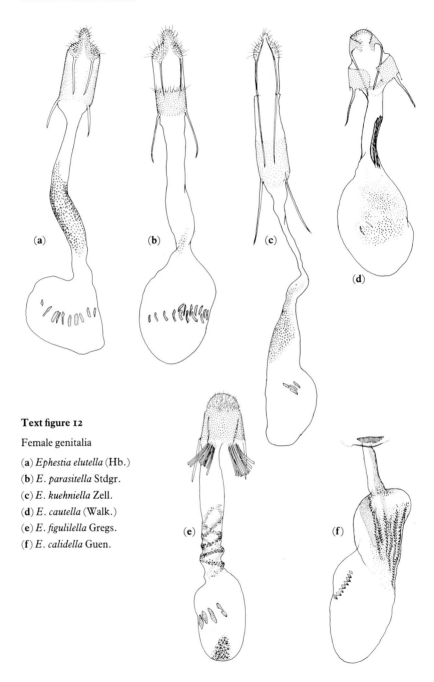

Text figure 12

Female genitalia

(a) *Ephestia elutella* (Hb.)
(b) *E. parasitella* Stdgr.
(c) *E. kuehniella* Zell.
(d) *E. cautella* (Walk.)
(e) *E. figulilella* Gregs.
(f) *E. calidella* Guen.

sprinkled dark fuscous; first line very slightly paler than ground-colour, dark-edged distally, the edging thicker towards costa, oblique to disc, then angled basad and strongly indented at dorsum; second line conspicuous only on account of its dark edging, serrate and strongly indented near costa; two transversely placed black discal dots often joined, a series of black crescentic interneural dots in termen; fringe whitish, fuscous at base and bisected by a faintly darker fuscous line. Hindwing whitish with fuscous veins, costa and termen; fringe white with faint fuscous line near base. The largest *Ephestia* species, forewing darker grey than in the other species, discal dots more conspicuous and second line characteristically deeply indented near costa; thickened dorsal border of valva in male with slightly recurved short tapered process at tip (text fig. 11c); ductus bursae of female narrow with band of minute spines near junction with bursa copulatrix (text fig.12c).

Flies from April to October in three overlapping broods; rests by day on inside walls, woodwork and windows of mills and warehouses, and flies at dusk and after dark. Seldom if ever reported out of doors in Britain and can survive the winter only in warmer parts of buildings.

LARVA Feeds principally on wheat-flour, but also on other stored food material and sometimes on dead insects. In mills, it is often a serious pest, clogging machinery with masses of silken, frass-filled galleries and webs.

DISTRIBUTION Found in flour-mills, granaries and bakeries throughout the country, though apparently not recorded in Britain before 1887 (Beirne, 1952).

Ephestia cautella (Walker) Dried Currant Moth

Pl.8, figs 35,36; text figs 11d,12d

IMAGO Wingspan 15–22mm. Forewing almost unicolorous reddish grey or brownish, the two cross-lines very indistinctly indicated in whitish, the first rather broad, darkened along outer border, nearly straight, slightly oblique; two indistinct discal spots transversely placed. Hindwing light grey. Valve of male genitalia characteristically shaped, with short projection on dorsal margin just beyond half-way (text fig.11d); female with sclerotized band along side of ductus bursae, half length of corpus bursae (text fig.12d); corpus bursae lacking sclerotized patch. The genitalia are most like those of *E. calidella* Guenée, *q.v.*, but the male of that species has among other differences a larger, thicker thumb-like projection on dorsal margin of valva and in female the sclerotized band on side of ductus bursae is one-third or less length of bursa copulatrix, and bursa copulatrix has a large, sclerotized patch posteriorly.

Flies in May and June and sometimes as a partial second brood in late summer and autumn. Rests by day on walls and windows of warehouses containing dried fruit and other produce, and flies in the buildings after dark.

LARVA Feeds chiefly on dried fruit, but also on a variety of stored vegetable material, including grain, flour, nuts and seeds. The larva spins a dense silken, frass-filled gallery amongst the food material. Winter is passed as a full-fed larva in a silken cocoon amongst the food.

DISTRIBUTION Frequently introduced from abroad into dried-fruit warehouses in towns throughout England, Wales and Ireland, probably also Scotland. Jersey, one in 1964 (Long, 1967).

Ephestia figulilella Gregson   Raisin Moth

Pl.8, figs 33,34; text figs 11e,12e

IMAGO Wingspan 15–20mm. Forewing light yellowish grey, the markings usually very indistinct; first line extends almost vertically from dorsum then bends at three-quarters to meet costa obliquely. Hindwing whitish, in male with ochreous yellow hair-tuft at base. Valva of male genitalia with prominent process about three-times as long as broad, at about half-way (text fig.11e); ductus bursae of female with spiral band of minute spines (text fig.12e).

Flies mainly in July and August, but sometimes also as a second emergence in autumn. Occurs in warehouses, but has also been taken out of doors.

LARVA Feeds from September to June on dried fruits and meal.

DISTRIBUTION The least common of the British *Ephestia* species inhabiting warehouses. Reported from warehouses in the larger towns since the first record in 1875 (Beirne, 1952).

Ephestia calidella Guenée   Dried Fruit Moth

Pl.8, fig.37; text figs 11f,12f

IMAGO Wingspan 19–24mm. Forewing greyish fuscous or fuscous, irrorated with whitish and dark fuscous scales; first line straight from costa to dorsum, pale, broadly edged blackish; second line edged blackish on both sides; two dark, transversely placed discal spots sometimes united to form a streak. Hindwing whitish grey with darker termen. Valva of male genitalia with thick, oblique projection about twice as long as broad, beyond middle (text fig.11f); ductus bursae of female with sclerotized band about one-third length of bursa copulatrix, and with large sclerotized patch posteriorly on corpus bursae (text fig.12f). One of the larger species of *Ephestia*, matching *E. kuehniella* (Pl.8, fig.32) in size, but with genitalia most resembling those of *E. cautella*, *q.v.*

Flies chiefly in August and September.

LARVA Feeds from September to May on dried fruits, nuts, cork and sometimes on other dried plant and animal material.

DISTRIBUTION Like *E. figulilella*, an uncommon species in British warehouses, as far north as the Clyde.

Colour Plates

1 *Chilo phragmitella* (Hübner) ♂.
   Page 22
2 *Chilo phragmitella* (Hübner) ♀.
   Page 22
3 *Calamotropha paludella* (Hübner)
   ♂. *Page 24*
4 *Calamotropha paludella* (Hübner)
   ♀. *Page 24*
5 *Acigona cicatricella* (Hübner) ♂.
   Page 23
6 *Acigona cicatricella* (Hübner) ♀.
   Page 23
7 *Chrysoteuchia culmella* (Linnaeus).
   Page 24
8 *Chrysoteuchia culmella* (Linnaeus).
   Page 24
9 *Euchromius ocellea* (Haworth).
   Page 22
10 *Crambus pascuella* (Linnaeus) ♂.
   Page 25
11 *Crambus pascuella* (Linnaeus) ♀.
   Page 25
12 *Crambus leucoschalis* Hampson.
   Page 25
13 *Crambus silvella* (Hübner). *Page 26*
14 *Crambus uliginosellus* Zeller.
   Page 26
15 *Crambus ericella* (Hübner). *Page 27*
16 *Crambus pratella* (Linnaeus).
   Page 28
17 *Crambus lathoniellus* Zincken ♂.
   Page 28
18 *Crambus lathoniellus* Zincken ♀.
   Page 28
19 *Crambus lathoniellus* Zincken ♂
   (W. Ireland). *Page 28*
20 *Crambus hamella* (Thunberg).
   Page 27
21 *Crambus perlella* (Scopoli). *Page 29*
22 *Crambus perlella* (Scopoli)
   f. *warringtonellus* Stainton. *Page 29*
23 *Agriphila straminella* (Denis &
   Schiffermüller) ♂. *Page 30*
24 *Agriphila straminella* (Denis &
   Schiffermüller) ♀. *Page 30*
25 *Agriphila straminella* (Denis &
   Schiffermüller) (Shetland) ♀.
   Page 30
26 *Agriphila tristella* (Denis &
   Schiffermüller) ♂. *Page 30*
27 *Agriphila tristella* (Denis &
   Schiffermüller) ♀. *Page 30*
28 *Agriphila latistria* (Haworth).
   Page 32
29 *Agriphila selasella* (Hübner).
   Page 29
30 *Agriphila inquinatella* (Denis &
   Schiffermüller) ♂. *Page 31*
31 *Agriphila inquinatella* (Denis &
   Schiffermüller) ♀. *Page 31*
32 *Agriphila inquinatella* (Denis &
   Schiffermüller) ♂. *Page 31*
33 *Agriphila poliellus* (Treitschke).
   Page 32

PLATE 1

1 *Agriphila geniculea* (Haworth).
Page 32

2 *Agriphila geniculea* (Haworth).
Page 32

3 *Catoptria permutatella*
(Herrich-Schäffer). *Page 33*

4 *Catoptria osthelderi* (de Lattin).
Page 33

5 *Catoptria speculalis* Hübner.
Page 35

6 *Catoptria pinella* (Linnaeus).
Page 35

7 *Catoptria margaritella* (Denis &
Schiffermüller). *Page 36*

8 *Catoptria furcatellus* (Zetterstedt).
Page 36

9 *Catoptria falsella* (Denis &
Schiffermüller). *Page 36*

10 *Catoptria verellus* (Zincken).
Page 37

11 *Catoptria lythargyrella* (Hübner).
Page 37

12 *Chrysocrambus linetella* (Fabricius).
Page 38

13 *Chrysocrambus craterella* (Scopoli).
Page 38

14 *Thisanotia chrysonuchella* (Scopoli).
Page 40

15 *Pediasia fascelinella* (Hübner) ♂.
Page 40

16 *Pediasia fascelinella* (Hübner) ♀.
Page 40

17 *Pediasia contaminella* (Hübner).
Page 41

18 *Pediasia contaminella* (Hübner)
ab. *sticheli* Constant. *Page 41*

19 *Pediasia aridella* (Thunberg).
Page 41

20 *Pediasia aridella* (Thunberg).
Page 41

21 *Platytes alpinella* (Hübner).
Page 42

22 *Platytes cerussella* (Denis &
Schiffermüller) ♂. *Page 42*

23 *Platytes cerussella* (Denis &
Schiffermüller) ♀. *Page 42*

24 *Ancylolomia tentaculella* (Hübner).
Page 43

25 *Schoenobius gigantella* (Denis &
Schiffermüller) ♂. *Page 44*

26 *Schoenobius gigantella* (Denis &
Schiffermüller) ♀. *Page 44*

27 *Schoenobius gigantella* (Denis &
Schiffermüller) ♂. *Page 44*

28 *Donacaula forficella* (Thunberg) ♂.
Page 45

29 *Donacaula forficella* (Thunberg) ♀.
Page 45

30 *Donacaula mucronellus* (Denis &
Schiffermüller) ♂. *Page 45*

31 *Donacaula mucronellus* (Denis &
Schiffermüller) ♀. *Page 45*

PLATE 2

1 *Scoparia subfusca* Haworth.
   Page 46

2 *Scoparia subfusca* Haworth,
   f. *scotica* White. *Page 46*

3 *Scoparia subfusca* Haworth
   (Orkney). *Page 46*

4 *Scoparia basistrigalis* Knaggs.
   Page 48

5 *Scoparia pyralella* (Denis &
   Schiffermüller). *Page 46*

6 *Scoparia pyralella* (Denis &
   Schiffermüller) f. *ingratella* Zeller,
   Knaggs *nec* Zeller. *Page 46*

7 *Scoparia pyralella* (Denis &
   Schiffermüller) f. *purbeckensis*
   Bankes. *Page 46*

8 *Scoparia ancipitella* (de la Harpe).
   Page 50

9 *Scoparia ambigualis* (Treitschke).
   Page 47

10 *Scoparia ambigualis* (Treitschke)
   f. *atomalis* Stainton. *Page 47*

11 *Scoparia ambigualis* (Treitschke)
   (Malham Tarn). *Page 47*

12 *Eudonia pallida* (Curtis). *Page 51*

13 *Eudonia alpina* (Curtis). *Page 51*

14 *Dipleurina lacustrata* (Panzer).
   Page 50

15 *Eudonia murana* (Curtis). *Page 52*

16 *Eudonia truncicolella* (Stainton).
   Page 53

17 *Eudonia lineola* (Curtis). *Page 53*

18 *Eudonia angustea* (Curtis). *Page 54*

19 *Eudonia delunella* (Stainton).
   Page 54

20 *Eudonia mercurella* (Linnaeus).
   Page 55

21 *Eudonia mercurella* (Linnaeus)
   f. loc. *portlandica* Curtis. *Page 55*

22 *Acentria ephemerella* (Denis &
   Schiffermüller) ♂ Water Veneer.
   Page 62

23 *Elophila nymphaeata* (Linnaeus) ♂
   Brown China-mark. *Page 55*

24 *Elophila nymphaeata* (Linnaeus) ♀
   Brown China-mark. *Page 55*

25 *Elophila nymphaeata* (Linnaeus) ♂
   Brown China-mark (New Forest).
   Page 55

26 *Elophila difflualis* (Snellen) ♂.
   Page 56

27 *Elophila difflualis* (Snellen) ♀.
   Page 56

28 *Elophila melagynalis* (Agassiz) ♂.
   Page 57

29 *Elophila melagynalis* (Agassiz) ♀.
   Page 57

30 *Elophila manilensis* (Hampson) ♂.
   Page 57

31 *Elophila manilensis* (Hampson) ♀.
   Page 57

32 *Nymphula stagnata* (Donovan)
   Beautiful China-mark. *Page 58*

33 *Nymphula stagnata* (Donovan)
   Beautiful China-mark. *Page 58*

34 *Parapoynx stratiotata* (Linnaeus) ♂
   Ringed China-mark. *Page 58*

35 *Parapoynx stratiotata* (Linnaeus) ♀
   Ringed China-mark. *Page 58*

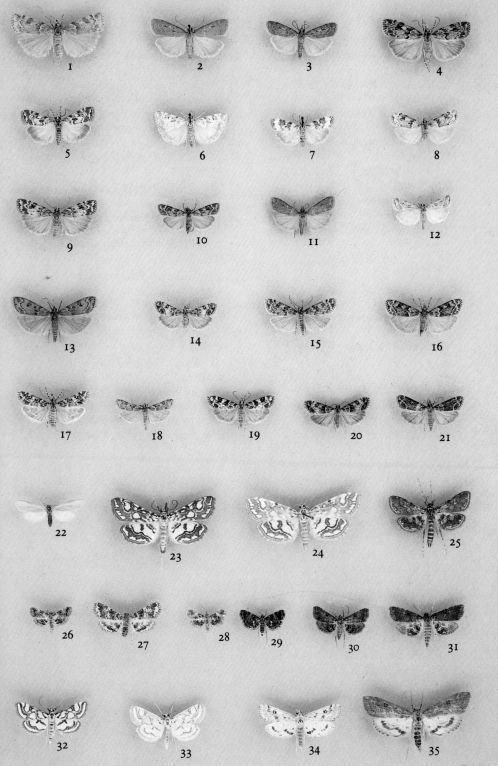

PLATE 3

PLATE 4

1 *Eurrhypara hortulata* (Linnaeus)
   Small Magpie. *Page 75*
2 *Perinephela lancealis* (Denis &
   Schiffermüller) ♂. *Page 75*
3 *Perinephela lancealis* (Denis &
   Schiffermüller) ♀. *Page 75*
4 *Phlyctaenia coronata* (Hufnagel).
   Page 76
5 *Mutuuraia terrealis* (Treitschke).
   Page 78
6 *Phlyctaenia perlucidalis* (Hübner).
   Page 76
7 *Anania funebris* (Ström). *Page 78*
8 *Anania funebris* (Ström) (Burren,
   W. Ireland). *Page 78*
9 *Anania verbascalis* (Denis &
   Schiffermüller). *Page 79*
10 *Phlyctaenia stachydalis* (Germar).
   Page 77
11 *Psammotis pulveralis* (Hübner).
   Page 79
12 *Ebulea crocealis* (Hübner). *Page 80*
13 *Opsibotys fuscalis* (Denis &
   Schiffermüller). *Page 81*
14 *Nascia cilialis* (Hübner). *Page 81*
15 *Udea lutealis* (Hübner). *Page 82*
16 *Udea fulvalis* (Hübner). *Page 82*
17 *Udea prunalis* (Denis &
   Schiffermüller). *Page 83*
18 *Udea decrepitalis*
   (Herrich-Schäffer). *Page 83*
19 *Udea olivalis* (Denis &
   Schiffermüller). *Page 84*
20 *Udea uliginosalis* (Stephens) ♂.
   Page 84
21 *Udea uliginosalis* (Stephens) ♀.
   Page 84
22 *Udea alpinalis* (Denis &
   Schiffermüller). *Page 85*
23 *Udea ferrugalis* (Hübner). *Page 85*

24 *Mecyna flavalis* (Denis &
   Schiffermüller)
   subsp. *flaviculalis* Caradja. *Page 86*
25 *Mecyna asinalis* (Hübner). *Page 86*
26 *Nomophila noctuella* (Denis &
   Schiffermüller) Rush Veneer.
   Page 87
27 *Nomophila nearctica* Munroe.
   Page 87
28 *Dolicharthria punctalis* (Denis &
   Schiffermüller) ♂. *Page 88*
29 *Antigastra catalaunalis*
   (Duponchel). *Page 88*
30 *Maruca testulalis* (Geyer)
   Mung Moth. *Page 89*
31 *Diasemia reticularis* (Linnaeus).
   Page 89
32 *Diasemiopsis ramburialis*
   (Duponchel). *Page 90*
33 *Hymenia recurvalis* (Fabricius).
   Page 90

PLATE 5

1 *Pleuroptya ruralis* (Scopoli) Mother of Pearl. *Page 91*

2 *Herpetogramma centrostrigalis* (Stephens). (Holotype). *Page 91*

3 *Herpetogramma aegrotalis* (Zeller). *Page 91*

4 *Palpita unionalis* (Hübner). *Page 92*

5 *Diaphania hyalinata* (Linnaeus), Melon-worm. *Page 92*

6 *Agrotera nemoralis* (Scopoli). *Page 92*

7 *Leucinodes vagans* (Tutt). *Page 93*

8 *Sceliodes laisalis* (Walker). *Page 93*

9 *Hypsopygia costalis* (Fabricius) Gold Triangle. *Page 94*

10 *Hypsopygia costalis* (Fabricius) ab. Gold Triangle. *Page 94*

11 *Synaphe punctalis* (Fabricius) ♂. *Page 94*

12 *Synaphe punctalis* (Fabricius) ♀. *Page 94*

13 *Orthopygia glaucinalis* (Linnaeus). *Page 95*

14 *Pyralis lienigialis* (Zeller). *Page 95*

15 *Pyralis farinalis* (Linnaeus) Meal Moth. *Page 96*

16 *Pyralis manihotalis* Guenée. *Page 96*

17 *Pyralis pictalis* (Curtis) Painted Meal Moth. *Page 97*

18 *Aglossa caprealis* (Hübner) Small Tabby. *Page 97*

19 *Aglossa pinguinalis* (Linnaeus) ♂ Large Tabby. *Page 97*

20 *Aglossa pinguinalis* (Linnaeus) ♀ Large Tabby. *Page 97*

21 *Aglossa dimidiata* (Haworth) Tea Tabby. *Page 98*

22 *Aglossa ocellalis* Lederer. *Page 98*

23 *Endotricha flammealis* (Denis & Schiffermüller). *Page 98*

24 *Endotricha flammealis* (Denis & Schiffermüller) (melanic form). *Page 98*

25 *Galleria mellonella* (Linnaeus) ♂ Wax Moth. *Page 99*

26 *Galleria mellonella* (Linnaeus) ♀ Wax Moth. *Page 99*

27 *Achroia grisella* (Fabricius) Lesser Wax Moth. *Page 100*

28 *Aphomia sociella* (Linnaeus) ♂ Bee Moth. *Page 101*

29 *Aphomia sociella* (Linnaeus) ♀ Bee Moth. *Page 101*

30 *Corcyra cephalonica* (Stainton) Rice Moth. *Page 100*

31 *Melissoblaptes zelleri* (de Joannis) ♀. *Page 101*

32 *Melissoblaptes zelleri* (de Joannis) (fuscous form) ♀. *Page 101*

33 *Melissoblaptes zelleri* (de Joannis) ♂. *Page 101*

PLATE 6

1 *Paralipsa gularis* (Zeller) ♂ Stored Nut Moth. *Page 102*

2 *Paralipsa gularis* (Zeller) ♀ Stored Nut Moth. *Page 102*

3 *Arenipses sabella* Hampson. *Page 102*

4 *Anerastia lotella* (Hübner). *Page 103*

5 *Cryptoblabes bistriga* (Haworth). *Page 103*

6 *Cryptoblabes gnidiella* (Millière). *Page 104*

7 *Oncocera semirubella* (Scopoli). *Page 107*

8 *Oncocera semirubella* (Scopoli) (plain form). *Page 107*

9 *Pempelia palumbella* (Denis & Schiffermüller). *Page 108*

10 *Pempelia genistella* (Duponchel). *Page 108*

11 *Pempelia obductella* (Zeller). *Page 109*

12 *Pempelia formosa* (Haworth). *Page 110*

13 *Salebriopsis albicilla* (Herrich-Schäffer). *Page 104*

14 *Sciota hostilis* (Stephens). *Page 110*

15 *Selagia argyrella* (Denis & Schiffermüller). *Page 106*

16 *Epischnia bankesiella* Richardson. *Page 111*

17 *Phycita roborella* (Denis & Schiffermüller). *Page 107*

18 *Phycita roborella* (Denis & Schiffermüller) (fuscous form). *Page 107*

19 *Dioryctria abietella* (Denis & Schiffermüller). *Page 112*

20 *Dioryctria mutatella* Fuchs. *Page 113*

21 *Dioryctria schuetzeella* Fuchs. *Page 113*

22 *Hypochalcia ahenella* (Denis & Schiffermüller). *Page 111*

23 *Microthrix similella* (Zincken). *Page 106*

24 *Metriostola betulae* (Goeze). *Page 105*

25 *Pyla fusca* (Haworth). *Page 106*

26 *Pima boisduvaliella* (Guenée). *Page 114*

27 *Trachonitis cristella* (Hübner). *Page 105*

28 *Pempeliella ornatella* (Denis & Schiffermüller). *Page 117*

29 *Pempeliella diluta* (Haworth). *Page 115*

30 *Pempeliella diluta* (Haworth) (W. Ireland). *Page 115*

31 *Nephopterix angustella* (Hübner). *Page 115*

32 *Acrobasis consociella* (Hübner). *Page 118*

33 *Acrobasis tumidana* (Denis & Schiffermüller). *Page 117*

34 *Acrobasis repandana* (Fabricius). *Page 118*

PLATE 7

PLATE 8

References

AGASSIZ, D. J. L., 1978. Five introduced species, including one new to science, of china mark moths (Lepidoptera: Pyralidae) new to Britain. *Entomologist's Gaz.* **29**: 117–127.

———, 1981a. Further introduced china mark moths (Lepidoptera: Pyralidae) new to Britain. *Ibid.* **32**: 21–26.

———, 1981b. *A revised list of the Lepidoptera (moths and butterflies) of the Isles of Scilly*, 20 pp., 1 map. Isles of Scilly Museum Association [Publication No. 14].

———, 1982. *Parapoynx stagnalis* (Zeller) (Lepidoptera: Pyralidae): a correction. *Entomologist's Gaz.* **33**: 122

———, 1983. Microlepidoptera – a review of the year 1982. *Entomologist's Rec.J. Var.* **95**: 187–195.

———, 1984. Microlepidoptera – a review of the year 1983. *Ibid.* **96**: 245–258.

BAKER, C. R., 1976. Pest Species. *In* Heath J. (ed.), *The Moths and Butterflies of Great Britain and Ireland*, **1**: 71–91. London.

BARRETT, C. G., 1886. Occurrence of *Botys repandalis*, Schiff., in Britain. *Entomologist's mon. Mag.* **23**: 145.

———,1904. *The Lepidoptera of the British Islands*, **9**: 454 pp., pls 377–424 [Pyralidae: pp. 160–340, 410–448, pls 395–412, 420–424]; **10** [1904–05]: 384 pp., pls 425–469 [Pyralidae [1904]: pp. 1–152, pls 425–443]. London.

BEIRNE, B. P., 1952. *British pyralid and plume moths*, 208 pp., 16 pls, 189 figs. London.

BLAND, K. P., 1983. Notes on Scottish Microlepidoptera, 1982. *Entomologist's Rec.J.Var.* **95**: 183–184.

BLESZYŃSKI, S., 1965. *Microlepidoptera palaearctica I: Crambinae*, **1** [text]: xlvii, 553 pp., 3 maps, 368 text figs; **2** [plates]: 133 pls, 369 text figs. Vienna.

BOND, K. G. M., 1979. New Irish records of Microlepidoptera in August. *Entomologist's Rec.J.Var.* **91**: 27.

BRADLEY, J. D. & FLETCHER, D. S., 1979. *A recorder's log book or label list of British butterflies and moths* [vi], 136 pp. London.

———, 1983. *A recorder's log book or label list of British butterflies and moths, Index* (D. H. Hall-Smith): *Addenda and Corrigenda*, [vi], 59 pp. Leicestershire Museums Publication No. 41. Leicester.

BRETHERTON, R. F., 1962. *Diasemia ramburialis* Duponchel and *D. litterata* Scopoli in Britain. *Entomologist's Rec.J.Var.* **74**: 1–9.

CAPPS, H. W., 1966. Review of new world moths of genus *Euchromius* Guenée with descriptions of two new species (Lepidoptera: Crambidae). *Proc.U.S.natn. Mus.* **119**: 4.

CARTER, D. J., 1967. Notes on *Catoptria permutatellus* Herrich-Schäffer and two closely allied species new to the British list (Lepidoptera, Pyralidae: Crambinae). *Entomologist's Gaz.* **18**: 91–94.

CHALMERS-HUNT, J. M., 1952. *Chilo cicatricellus* Hübner confirmed as British. *Entomologist's Rec.J.Var.* **64**: 160–161.

———, 1968. *Maruca testulalis* (Geyer): 'The bean pod borer' (Lep.: Pyralidae) bred out at East Malling from french beans. *Ibid.* **80**: 242, pl. XIII.

———, 1970. The butterflies and moths of the Isle of Man. *Trans.Soc.Br.Ent.* **19** Pt. 1: 1–171.

——— & Tweedie, M. W. F., 1982. *Dioryctria schuetzeella* Fuchs, 1899: a pyralid moth new to Britain. *Entomologist's Rec.J.Var.* **94**: 1–3, figs 1–7.

CLUTTERBUCK, C. GRANVILLE, 1930. *Phlyctaenia fulvalis* Hb., a new British pyralid. *Entomologist* **63**: 85–86.

COCKAYNE, E. A. (ed.), 1953. Field notes. *Entomologist's Rec.J.Var.* **65**: 13–14.

CORLEY, M. F. V., 1985. *Pyralis lienigialis* (Zeller) in Oxfordshire and Wiltshire (Lepidoptera: Pyralidae). *Entomologist's Gaz.* **36**: 189–192.

DACIE, J., 1984a. *Maruca testulalis* Geyer in Surrey. *Entomologist's Rec.J.Var.* **96**: 28.

———, 1984b. *Daraba laisalis* in Surrey. *Ibid.* **96**: 123.

DE WORMS, C. G. M., 1979. *Maruca testulalis* Geyer (Lep.: Pyralidae) in the London area. *Ibid.* **91**: 286.

DURRANT, J. H. (ed.), 1919. Occurrence of *Areniphes* [*sic*] *sabella* Hmsn. (Galleriadae) in London. *Proc.ent.Soc.London* **1919**: xiv.

EDWARDS, T. G. & WAKELY, S., 1952. *Ancylolomia tentaculella* Hübn. in Kent. *Entomologist's Rec.J.Var.* **64**: 273–274.

EMMET, A. M. (ed.), 1979. *A field guide to the smaller British Lepidoptera*, 271 pp. London.

———, 1984. *Euzophera bigella* (Zeller) (Lepidoptera: Pyralidae): a second British record. *Entomologist's Gaz.* **35**: 154.

FASSNIDGE, W., 1928. Notes on *Myelois cirrigerella* Zinck. *Trans.Hamps.ent.Soc.* **4**: 34–36.

———, 1935a. *Cateremna terebrella* Zinck. in Buckinghamshire and Hampshire. *Entomologist's Rec.J.Var.* **47**: 51–52.

———, 1935b. *Myelois cirrigerella* Zinck. taken near Winchester, with notes on the larval habits. *Ibid.* **47**: 115.

FORD, L. T., 1936. *Melissoblaptes bipunctanus*, Zeller 1848. *Ibid.* **48**: 93–94.

———, 1957. A note on the ova and larva of *Heterographis oblitella* (Zeller) (Lep.: Phycitidae). *Entomologist's Gaz.* **8**: 27.

FOSTER, A. P., 1984. *Maruca testulalis* (Geyer) (Lep.: Pyralidae) in Cornwall. *Entomologist's Rec.J.Var.* **96**: 28.

FREEMAN, P. (ed.), 1980. *Common insect pests of stored food products*, British Museum (Natural History) Economic Series No. 15; edn 6, 69 pp., 130 figs. London.

FRYER, J. C. F., 1933. *Phlyctenia* [*sic*] *fulvalis* and *Crambus contaminellus* in southern England. *Entomologist* **66**: 265–267.

GARDINER, B. O. C., 1961. The Pyraloidea of Cambridgeshire and Huntingdonshire. *Entomologist's Gaz.* **12**: 173–192.

GOATER, B., 1974. *The butterflies and moths of Hampshire and the Isle of Wight*, 439 pp. Faringdon.

GRIFFITH, A. F., 1881. *Crambus verellus* at Cambridge. *Entomologist* **14**: 20.

HANCOCK, E. G., 1984. *Oligostigma bilinealis* Snellen (Lepidoptera: Pyralidae), a second British record. *Entomologist's Gaz.* **35**: 18.

HANNEMANN, H.-J., 1964. Kleinschmetterlinge oder Microlepidoptera II, Die Winkler (s.l.) (Cochylidae und Carposinidae). Die Zünslerartigen (Pyraloidea). *Tierwelt Dtl.* **50**: 78–376, 385–401, pls 4–22.

HEATH, J. & EMMET, A. M. (eds), in preparation. *The moths and butterflies of Great Britain and Ireland*, **6** (Alucitidae, Pyralidae and Pterophoridae). Colchester.

HOLST, P. L., 1978. Supplementary notes on *Pyrausta ostrinalis* Hb. *Lepidoptera Kbh.* (new series) **3**: 121–127.

HOWARTH, T. G., 1950. Notes on the life history of *Margaronia unionalis* Hbn. *Entomologist's Gaz.* **1**: 85–88, 2 figs.

HUGGINS, H. C., 1927. Variation in *Gymnancyla canella*. *Entomologist* **60**: 279–280.

———, 1929a. *Salebria obductella* in Kent. *Ibid.* **62**: 52–53.

———, 1929b. *Salebria obductella* (Lep. Pyralidae) as a resident species. *Ibid.* **62**: 193–195.

———, 1929c. The pupa of *Homaeosoma* [*sic*] *cretacella* Rössl. (=*senecionis* Vaughan). *Ibid.* **62**: 232–233.

———, 1954. Notes on Microlepidoptera [re *Crambus verellus* Zinck.] *Entomologist's Rec.J.Var.* **66**: 54–55.

———, 1959. A British capture of *Crambus leucoschalis* Hampson. *Entomologist* **92**: 176.

HULME, D. C., 1968. Pyralid and plume moths of Derbyshire. *Entomologist's Rec.J.Var.* **80**: 157–163, 221–227.

JACOBS, S. N. A., 1933. *Aphomia gularis* (Zell.) and other rare warehouse moths. *Entomologist* **66**: 195.

———, 1935. *Aphomia gularis* Zell. *Proc.Trans.S.Lond.ent.nat.Hist.Soc.* **1934–35**: 99–104.

JEWESS, P. J., 1982. *Udea decrepitalis* H.-S. (Lep.: Pyralidae) in Wales. *Entomologist's Rec.J.Var.* **94**: 121–122.

KLOET, G. S. & HINCKS, W. D., 1972. A check list of British insects: Lepidoptera (Edn 2). *Handbk Ident.Br. Insects* **11** (2), viii, 153 pp.

KLOTS, A. B., 1970. In Tuxen, S. L., *Taxonomist's glossary of genitalia in insects* (Edn 2), pp. 115–130. Copenhagen.

LEECH, J. H., 1886. *British pyralides, including the Pterophoridae*, viii, 122 pp., 18 pl. London.

LERAUT, P., 1980. *Liste systématique at synonymique des Lépidoptères de France, Belgique et Corse*, 334 pp. Paris.

———, 1983 (1984). Contribution à l'étude des Scopariinae. 4. Révision des types décrits de la région paléarctique occidentale, description de dix nouveaux taxa et ébauche d'une liste des espèces de cette région. (Lep. Crambidae). *Ibid.* **13**: 157, 192.

LHOMME, L., 1935–46. *Catalogue des Lépidoptères de France et de Belgique*, 2(1): 584 pp. Le Carriol.

LONG, R., 1967. The pyralid and plume moths of Jersey. *A.Bull.Soc.jersiaise* **19**: 225–232.

LORIMER, R. I., 1983. *The Lepidoptera of the Orkney Islands*, viii, 103 pp. Faringdon.

MASON, P. B., 1892. *Hercyna phrygialis* Hb., probably a British insect? *Entomologist's mon.Mag.* **28**: 264.

MERE, R. M., 1952. A pyrale new to Britain [re *Hymenia recurvalis* Fabr.]. *Entomologist's Gaz.* **3**: 56–57.

———, 1965. *Nephopterix albicilla* Herrich-Schäffer in Monmouthshire – a phycitid moth new to the British Isles. *Ibid.* **16**: 13–14, pl. I.

——— & BRADLEY, J. D., 1957. *Pyrausta perlucidalis* (Hübner), a pyralid new to the British Isles. *Ibid.* **8**: 162–166, pl. IV, text figs 1–4.

MEYRICK, E., 1928. *A revised handbook of British Lepidoptera*, vi, 914 pp. London.

MORLEY, A. M., 1953. *Ancylolomia tentaculella*. Hüb.: a belated record. *Entomologist's Rec.J.Var.* **65**: 148.

MUNROE, E. G., 1973. A supposedly cosmopolitan insect: the celery webworm and allies, genus *Nomophila* Hübner (Lepidoptera: Pyralidae: Pyraustinae). *Can.Ent.* **105**: 177–216, figs 1–47.

PELHAM-CLINTON, E. C., 1984. A British specimen of *Nomophila nearctica* Munroe (Lepidoptera: Pyralidae). *Entomologist's Gaz.* **35**: 155–156.

PIERCE, F. N., 1937. The British species of the *nimbella* group of the genus *Homoeosoma* (Lep., Pyralidae). *Entomologist* **70**: 97–100, pl. III.

——— & METCALFE, J. W., 1938. *The genitalia of the British pyrales, with the deltoids and plumes*, xiii, 69 pp., 29 pls. Oundle. (Reprint, 1968. Hampton.)

POTTER, C., 1930. A new moth introduced into England: *Mussidia nigrivenella* Ragonot (Phycitidae). *Entomologist's mon.Mag.* **66**: 258.

RILEY, N. D., 1930. *Phlyctaenia fulvalis* Hb. *Entomologist* **63**: 137, pl. III, fig. 6.

ROCHE, J., 1963. *Catoptria osthelderi* de Lattin, 1950: in Britain. *Entomologist's Rec.J.Var.* **75**: 141.

ROESLER, R. U., 1973. *Microlepidoptera Palaearctica IV: Phycitinae*, 1 [text]: xvi, 752 pp., 302 text figs; 2 [plates]: 137 pp., 170 pls. Vienna.

SHAFFER, M., 1966. Some changes in the nomenclature of British Lepidoptera, part 2, Pyraloidea [re the British species of *Dioryctria* Zeller, p. 22]. *Entomologist's Gaz.* **17**: 19–26.

———, 1968a. Illustrated notes on *Parapoynx obscuralis* (Grote) and *Hellula undalis* (Fabricius), two species new to the British list (Lepidoptera: Pyralidae). *Ibid.* **19**: 107–112, pl. 7, text figs 1–7.

———, 1968b. Illustrated notes on *Synclita obliteralis* (Walker) and *Euzophera bigella* (Zeller), two species new to the British list (Lepidoptera: Pyralidae). *Ibid.* **19**: 155–158, pl. 8, text figs 1–7.

SKINNER, B., 1982. The history of *Euchromius ocellea* (Haworth) (Lep.: Pyralidae) in Britain. *Entomologist's Rec.J.Var.* **94**: 139–140.

SPEIDEL, W., 1984. Revision der Acentropinae des palaearktischen Faunenge-bietetes (Lepidoptera, Crambidae) [re *Acentria ephemerella* ([Den. & Schiff.])]. *Neue ent. Nachr.* **12**: 70.

STAINTON, H. T. (ed.), 1855. New British species since 1835 [re *Chilo cicatricellus* Hübner]. *Entomologist's Annu.* **1855**: 23–24.

STEPHENS, J. F., 1834. *Illustrations of British entomology.* (Haustellata) **4**: 436 pp., 9 pls. London.

TUTT, J. W., 1890a. Current notes [re exhibition of *Botys mutualis*, Zell. = *Herpeto-gramma aegrotalis* (Zeller)] *Entomologist's Rec J. Var.* **1**: 33.

——, 1890b. *Aphytoceros vagans* (mihi) a species new to science *Ibid.* **1**: 203.

—— (ed.), 1893. *Aphytoceros vagans* [illustrated]. *Ibid.* **4**: Pl. C, fig. 6, facing p. 29.

VINE HALL, J. H., 1959. The genus *Crambus* F. (Lep., Pyralidae) in south Westmorland and Furness. *Entomologist's Gaz.* **10**: 141–144.

——, 1966. The *Scoparia* Haworth species of the Lake Counties (Lep.: Pyrali-dae). *Ibid.* **17**: 73–78.

WAKELY, S., 1933. *Aphomia gularis* (Zeller) in Britain. *Entomologist* **66**: 99, Pl. I, figs 8,9.

——, 1935. *Myelois neophanes* Durrant. *Ibid.* **68**: 137–138.

——, 1937. *Cryptoblabes gnidiella* Millière. *Ibid.* **70**: 71

WATERSTON, A. R., 1981. Present knowledge of the non-marine invertebrate fauna of the Outer Hebrides. *Proc.R.Soc.Edinb.* (B) **79**: 215–321.

WEBB, K. F., 1984. *Daraba laisalis* in Bedfordshire. *Entomologist's Rec.J.Var.* **96**: 123.

WHALLEY, P. E. S., 1959. The British species of the genus *Chrysocrambus* Blesz. (*Crambus* auctt.) Lepid. Pyralidae. *Entomologist* **92**: 179–184, Pl. IX, figs 1–4.

——, 1961. *Cadra woodiella* R. & T. a synonym of *C. parasitella* Staud. (Lep.: Pyralidae). *Entomologist's Gaz.* **12**: 113.

——, 1963. *Euzophera osseatella* Treitschke (Lep., Phycitinae) on potatoes imported from Egypt to Scotland. *Ibid.* **14**: 100.

—— & TWEEDIE, M. W. F., 1963. A revision of the British Scoparias (Lep.: Pyralidae). *Ibid.* **14**: 81–98.

WOOD, W., 1839. *Index entomologicus*, xii, 266 pp., 54 pls. London.

WORMELL, P., 1983. Lepidoptera of the Inner Hebrides. *Proc.R.Soc.Edinb.* (B) **83**: 531–546.

ZELLER, P. C., 1852. *Lepidoptera Microptera*, 120 pp. Stockholm.

Glossary

aedeagus – the tubular portion of the male penis, which contains within it the delicate, eversible *vesica* bearing spiny *cornuti*; the whole structure is of great diagnostic value in separating closely related species.

anastomose – to join together, as in two veins of the wing which unite having arisen separately.

antemedian – before the middle (of the wing).

antrum – the sclerotized anterior portion of the *ductus bursae*.

apex – the tip of the wing, junction of *costa* and *termen*.

auctt. (auctorum) – of other authors.

basad – directed towards the base (of the wing).

basal – near or at the base (of the wing).

bursa copulatrix – the bulbous base (anterior end) of the female genitalia, often adorned with one or more patches (*signa*) of minute thorn-like processes the details of which are of diagnostic value.

cell – the more or less oval-shaped area of the wing, from the base to about the middle, which is enclosed by veins.

chitin – the hard material of which an insect's skeleton is constructed.

cilium (pl. cilia) – (i) fine hair-like processes on the antenna.
(ii) fine hair-like specialised scales which form a fringe along the termen of fore- and hindwing.

ciliate – bearing *cilia*.

claviform stigma – a club-shaped mark arising from the first line above the dorsum, and pointing towards the middle of the wing. A rather uncommon feature in the Pyralidae (*Scoparia*, etc.)

concolorous – of the same colour.

cornutus (pl. cornuti) – spiny ornamentations of the male *vesica* (see *aedeagus*).

costa – the leading edge of the wing.

cross-lines – transverse linear markings characteristic of the wings of many Lepidoptera. In the Pyralidae, the *first line* (antemedial fascia) is situated before the middle of the forewing, the *second line* (postmedial fascia) a short distance beyond the middle, and the *subterminal line* (subterminal fascia) lies between the second line and the termen. In the hindwing, there is usually only one distinct line, if any; it lies beyond the middle and is referred to here as the *postmedian line* (postmedial fascia).

Numerous terms are available to describe the course of these cross-lines. Among them:

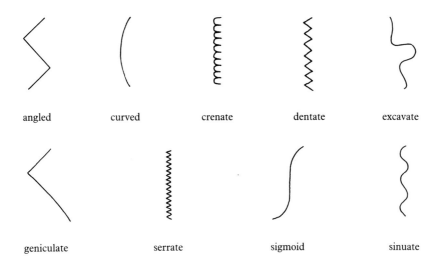

angled   curved   crenate   dentate   excavate

geniculate   serrate   sigmoid   sinuate

cucullus – the upper portion of the valva of the male genitalia; a term most usefully applied when this region is separated from the rest of the valva by a distinct constriction.

disc – the outer part of the *cell*.

discal spot – one or two (disco-cellular) spots, darker or paler than the surrounding area, which lie either within the boundary of the reniform stigma or in the region of the disc characteristically occupied by the reniform stigma.

disjunct – just separate from an adjacent structure or marking.

distad – directed towards the tip (of the wing).

distal – nearer the tip (of the wing).

diverticulum – a sac-like or tubular side-branch of a duct or tube.

dorsum – the trailing edge of the wing.

ductus bursae – the tube leading from the *bursa copulatrix* to the female genital aperture (*ostium*).

facial cone – a scale-covered conical projection on the front of the head.

fascia – a transverse band.

ferruginous – rust-coloured.

first line – see *cross-lines*.

fringe – the border of specialised scales (*cilia*) along the *termen* of the wing.

fuscous – greyish brown.

gnathos – the hinged, chitinised ventral lip of the armature surrounding the anus of the male.

hyaline – transparent and unscaled.

indentated – bearing a notch.

iridescence – a rainbow-coloured reflection from a surface the structure of which causes the refraction of light.

irrorate – finely and closely speckled.

interneural – between the veins.

juxta – a ring or tube of chitin, often heavily sclerotized, which forms part of the male genitalia and through which the penis is protruded.

labial palp – one of a pair of sensory appendages on the head adjacent to the mouthparts.

median – across the middle.

nacreous – pearly.

neural – referring to the veins of the wing.

obsolete – so weakly developed as to be hardly detectable.

orbicular stigma – a usually circular mark (antemedial spot) in the cell of the forewing, between the *first line* and the *reniform stigma*.

ostial plate – a flattened chitinous plate surrounding the opening of the female genital tract.

ostium – the external opening of the female genital tract.

pecten – the cubital pecten: a ridge of fine projecting hair-like scales along one of the veins of a wing.

porrected – directed forwards (labial palps).

postmedian – beyond the middle (of the wing).

reniform stigma – a typically kidney-shaped mark (disco-cellular spot) in the outermost region of the *cell* of the forewing. Its outline is usually unclear in Pyralidae (see *discal spot*).

reticulated – forming a fine network (of markings).

sclerotized – composed of a strong thick layer or band of dark-coloured *chitin*.

second line – see *cross-lines*.

seta (pl. **setae**) – a stiff bristle.

setose – bearing stiff bristles, *setae*

sibling – strictly, a brother or sister; hence, referring to two extremely closely related species.

signum (pl. **signa**) – patches of minute thorn-like processes on *bursa copulatrix*.

squamous – scaly.

stria – a fine streak.

strigulate – covered with fine, short transverse streaks.

subbasal – adjacent to the base (of the wing).

subcostal – adjacent to the *costa* (of the wing).

subdorsal – adjacent to the *dorsum* (of the wing).

subterminal line – see *cross-lines*.

suffused – delicately shaded as by a cloud.

tegumen – a chitinised ring close to the tip of the male abdomen, to which the penis is articulated.

termen – the outer margin of the wing.

tornus – the angle between the *termen* and *dorsum*.

tracheal gills – expanded membranous outgrowths of the bodies of certain aquatic larvae, richly supplied with branches of the insect's breathing apparatus, tracheae, which facilitate gas exchange in water.

trapezoidal – describing a four-sided marking, two sides of which are parallel.

tympanal organs – structures which are sensitive to sound; situated in pyralids at the base of the abdomen.

uncus – the hinged, chitinised dorsal lip of the armature surrounding the anus of the male (cf. *gnathos*).

valva (pl. **valvae**) – one of the paired clasping organs of the male genitalia.

vesica – see *aedeagus*.

Index to Food Plants
and other Larval Foodstuffs

Index to Insect Names

Principal entries are given in **bold type**. Plate references are shown as (*frontis*: 12; Pl. 7:19). The index includes references to figures in the text, as 49 (text fig. 5).

171